Laboratory Manual
to accompany

FUNDAMENTALS OF HVAC/R

Carter Stanfield
Athens Technical College

David Skaves
Maine Maritime Academy

Prepared by

David Skaves
Maine Maritime Academy

Prentice Hall
Upper Saddle River, New Jersey
Columbus, Ohio

Editor in Chief: Vernon Anthony
Acquisitions Editor: Eric Krassow
Editorial Assistant: Sonya Kottcamp
Project Manager: Louise Sette
Operations Supervisor: Laura Weaver
Art Director: Diane Ernsberger
Cover Designer: Bryan Huber
Cover Image: Getty One
Director of Marketing: David Gesell
Marketing Manager: Derril Trakalo
Senior Marketing Coordinator: Alicia Wozniak
Copyeditor: Carol Mohr

This book was set in 10/12 Times by Aptara, Inc., and was printed and bound by Bind-Rite Graphics. The cover was printed by Coral Graphic Services, Inc.

Pearson Prentice Hall™ is a trademark of Pearson Education, Inc.
Pearson® is a registered trademark of Pearson plc
Prentice Hall® is a registered trademark of Pearson Education, Inc.

Pearson Education Ltd., London
Pearson Education Singapore Pte. Ltd.
Pearson Education Canada, Inc.
Pearson Education—Japan

Pearson Education Australia Pty. Limited
Pearson Education North Asia Ltd., Hong Kong
Pearson Educación de Mexico, S.A. de C.V.
Pearson Education Malaysia Pte. Ltd.

Prentice Hall
is an imprint of

www.pearsonhighered.com

10 9 8 7 6 5 4 3 2 1
ISBN-13: 978-0-13-222410-9
ISBN-10: 0-13-222410-0

INTRODUCTION

This Laboratory Manual is designed to accompany the *Fundamentals of HVAC/R* textbook. Each laboratory worksheet is referenced back to this primary text. The laboratory worksheets can be completed in any order, however they have been laid out in a systematic format. The Fundamentals section provides the student with the basic skills for using typical refrigeration equipment. It is recommended that for introductory students this section be started first. The Properties section complements the Fundamentals section to reinforce the basic principles of fluid behavior based upon practical application. The Properties section may be used in parallel with the Fundamentals section as best suits the program needs. The Refrigeration section assumes a minimum level of refrigeration equipment knowledge and this section should not be started until students are thoroughly familiar with refrigeration testing and measuring equipment. The Accessories, Controls, and Electricity sections can be used in parallel with the Refrigeration section as best fits the goals for the program. The additional exercises on maintenance and heating are available for those programs that cover a broader range of topics than just air conditioning and refrigeration systems. Each laboratory worksheet's title reflects the topic of coverage making it clear as to what is contained in the exercise. This will help to determine how the exercise best fits into the program.

CONTENTS

MATERIAL SAFETY DATA SHEETS (MSDSs)

LABORATORY OBJECTIVE

The student will demonstrate an understanding of the information provided on Material Safety Data Sheets for one of the refrigerant types to be used in the refrigeration laboratory.

LABORATORY NOTES

This lab exercise should always be the first one performed at the beginning of the course. Students should be introduced to Material Safety Data Sheets (MSDSs) on the first day of lab.

MSDSs are required by law and have important information listed in specific areas so that they are easily read by emergency personnel.

You should read MSDSs on any material before you use it so you know how to use it properly and safely as well as knowing what to do if there is an accident involving the material.

FUNDAMENTALS OF HVAC/R TEXT REFERENCE
Unit 3

Required Tools and Equipment

Refrigerant Material Safety Data Sheet	Unit 3

SAFETY REQUIREMENTS
None

PROCEDURE

STEP 1. Review a copy of a refrigerant MSDS supplied by the Lab Instructor and answer the following:

A. What are the first aid measures for inhalation?

_____

_____

1

B. What are the first aid measures for refrigerant exposure to the eyes?

_____

_____

C. What are the first aid measures for refrigerant exposure to the skin?

_____

_____

D. What are the first aid measures for refrigerant ingestion?

_____

_____

E. What type of protective clothing is worn for the eyes and face?

_____

_____

F. What type of protective clothing is worn for the hands, arms, and body?

_____

_____

QUESTIONS

To get help in answering some of the following questions, refer to the *Fundamentals of HVAC/R* text Unit 3.

(Circle the letter that indicates the correct answer.)

1. If an area on an MSDS does not apply to a product:
   A. it should be left blank.
   B. it should be filed in with the refrigerant type.
   C. there should be a line through it.
   D. it should be marked as non-applicable.

2. The section of an MSDS that provides the properties of the material, such as boiling point, vapor pressure, and so on, is:
   A. Section I.
   B. Section II.
   C. Section III.
   D. Section IV.

3. If you do not follow the instructions for proper use and handling of a material as listed on an MSDS:
   A. you could be injured.
   B. a customer could be injured.
   C. you could be fired.
   D. All of the above are correct.

4. Section I of the MSDS provides:
   A. the manufacturer's name and address.
   B. the fire and explosion hazard data.
   C. hazardous ingredients/identity information.
   D. All of the above.

5. The HMIS information tells health care workers a relative number according to how significantly the material will affect health, how reactive it is, and its flammability.
   A. True.
   B. False.

DISPOSABLE REFRIGERANT CYLINDERS

LABORATORY OBJECTIVE

The student will demonstrate an understanding of disposable refrigerant cylinder types and handling procedures.

LABORATORY NOTES

For this lab exercise there should be a number of different disposable refrigerant cylinders for the students to inspect. Empty cylinders are acceptable as the student will not be removing any refrigerant from the cylinder. It would be preferable to also have cylinders of different sizes.

FUNDAMENTALS OF HVAC/R TEXT REFERENCE

Unit 26

Required Tools and Equipment

Disposable refrigerant cylinders of different sizes and types	Unit 26

SAFETY REQUIREMENTS

Caution!!! Do not open the valves on the refrigeration cylinders to avoid possible injury due to skin contact with refrigerant. Also, it is illegal to intentionally vent refrigerants composed of CFCs and HCFCs.

PROCEDURE

STEP 1. In 1990 the Congress of the United States passed a series of amendments to the Clean Air Act that greatly affected the refrigeration and air conditioning industry. The Act establishes a set of standards and requirements for the use and disposal of certain common refrigerants containing chlorine.

Review the Information on the Clean Air Act from the *Fundamentals of HVAC/R* text Unit 26 and answer the following:

A. Can you knowingly release CFCs, HCFCs, or HFCs while repairing appliances?

_____

_____

B. When did it become illegal to vent CFCs and HCFCs?

_____

_____

C. Do you need to be certified to service, maintain, or dispose of appliances containing refrigerants? If yes, when did this become mandatory?

_____

_____

D. How much can you or your company be fined per day for violating Section 608 of the Federal Clean Air Act?

_____

_____

E. How much is the bounty for turning someone in?

_____

_____

STEP 2. Locate a *disposable* refrigerant cylinder and examine it carefully so that you may complete the following exercise.

Refer to the *Fundamentals of HVAC/R* text Unit 26 for additional information.

A. From your cylinder, write down the refrigerant type, chemical designation of the refrigerant, cylinder color code, refrigerant boiling temperature, normal discharge (head) pressure, normal suction pressure, and latent heat value.

_____

_____

_____

_____

B. Give the refrigerant toxicity, flammability, corrosive tendency, and rated refrigerant weight when the cylinder is full.

_____

_____

_____

C. What is the maximum temperature that this cylinder can be exposed to?

_____

_____

D. How often must this cylinder be checked (DOT regulation)?

_____

_____

E. Can you remove both liquid and vapor refrigerant from this cylinder? If yes, then describe how you would remove vapor and how you would remove liquid refrigerant.

_____

_____

F. Can you reuse this cylinder in an emergency?

_____

_____

G. What type of protection against bursting due to excessive pressure does the cylinder have?

_____

_____

H. Draw a rough sketch of the protection device in the space provided below.

QUESTIONS

To get help in answering some of the following questions, refer to the *Fundamentals of HVAC/R* text Unit 26.

(Circle the letter that indicates the correct answer.)

1. A disposable cylinder:
   A. can be refilled *only* with the proper type of refrigerant.
   B. can never be refilled.
   C. is always color coded yellow and gray.
   D. Both B and C are correct.

2. Cylinders over 4.5 in in diameter and over 12 in long:
   A. will always be disposable cylinders.
   B. will always be recovery cylinders.
   C. cannot be used for refrigerant purposes.
   D. must have some type of pressure relief device.

3. When transporting a refrigerant cylinder:
   A. it should be properly secured in an upright position.
   B. it should be properly secured laying down.
   C. it should be properly secured in an inverted position.
   D. All of the above are correct.

4. Refrigerant cylinder temperatures should not exceed:
   A. 92.5°F.
   B. 99.8°F.
   C. −25°F.
   D. 125°F.

5. The cardboard boxes that contain new refrigerant cylinders:
   A. can be used to prop the bottle upright when charging.
   B. may contain important safety information.
   C. may need to be kept on the job site.
   D. Both B and C are correct.

6. To charge a liquid using a disposable cylinder:
   A. it must be heated.
   B. it must be cooled.
   C. it must be shaken.
   D. it must be inverted.

7. The burst disk on a disposable refrigerant cylinder.
   A. will always reseat.
   B. will completely drain the cylinder.
   C. will only drain the liquid refrigerant.
   D. acts like a relief valve.

REFILLABLE REFRIGERANT RECOVERY CYLINDERS

LABORATORY OBJECTIVE

The student will demonstrate an understanding of refillable refrigerant recovery cylinder types and handling procedures.

LABORATORY NOTES

For this lab exercise there should be a refillable refrigerant recovery cylinder for the students to inspect. Empty cylinders are acceptable as the student will not be removing any refrigerant from the cylinder.

FUNDAMENTALS OF HVAC/R **TEXT REFERENCE**

Unit 26

Required Tools and Equipment

Refillable refrigerant recovery cylinder	Unit 26

SAFETY REQUIREMENTS

Caution!!! Do not open the valves on the refrigeration cylinders to avoid possible injury due to skin contact with refrigerant. Also it is illegal to intentionally vent refrigerants composed of CFCs and HCFCs.

PROCEDURE

STEP 1. Locate a refillable refrigerant recovery cylinder and examine it carefully so that you may complete the following exercise.

Refer to the *Fundamentals of HVAC/R* text Unit 26 for additional information.

    A. Sketch a cross sectional view of the cylinder including fittings and valves in the space provided below.

B. Explain how can you remove either liquid or vapor from this cylinder.

_____

_____

C. What is the cylinder weight empty (T.W.—Tare Weight)?

_____

_____

D. What is the maximum refrigerant weight that the cylinder can hold?

_____

_____

E. What is the weight of a full cylinder including the metal?

_____

_____

F. What type of protection against bursting due to excessive pressure does the cylinder have?

_____

_____

G. How often must this cylinder be inspected?

_____

_____

H. What type of refrigerant can be stored in this cylinder?

_____

_____

QUESTIONS

To get help in answering some of the following questions, refer to the *Fundamentals of HVAC/R* text Unit 26.

(Circle the letter that indicates the correct answer.)

1. A refillable refrigerant recovery cylinder:
    A. can be refilled *only* with the proper type of refrigerant.
    B. can never be refilled.
    C. is always color coded yellow and gray.
    D. Both A and C are correct.

2. A refillable refrigerant recovery cylinder must meet DOT approval:
    A. True.
    B. False.

REFRIGERANT GAUGE MANIFOLDS

LABORATORY OBJECTIVE

The student will demonstrate how to properly use a refrigerant gauge manifold.

LABORATORY NOTES

For this lab exercise there should be a typical gauge manifold available for students to inspect.

FUNDAMENTALS OF HVAC/R TEXT REFERENCE

Units 12 and 25

Required Tools and Equipment

Gauge manifold	Units 12 and 25

SAFETY REQUIREMENTS

None

PROCEDURE

STEP 1. Locate a refrigerant gauge manifold and examine it carefully so that you may complete the following exercise.

Refer to the *Fundamentals of HVAC/R* text Unit 12 for additional information.

A. Identify the two gauges that are on the manifold including the minimum and maximum readings.

_____

_____

B. What is meant by compound gauge?

_____

_____

C. What is the range of the pressure scale on the compound gauge?

_____

_____

D. List three operations that can be performed with a gauge manifold.

_____

_____

_____

E. What do the colored scales in the center of the gauge indicate?

_____

_____

F. Sketch a cross sectional view of a gauge manifold, labeling all parts and connections in the space provided below.

QUESTIONS

To get help in answering some of the following questions, refer to the *Fundamentals of HVAC/R* text Units 12 and 25.

(Circle the letter that indicates the correct answer.)

1. When not in use:
   A. gauge manifolds should be stored in a plastic wrap cloth.
   B. gauge manifolds should be press charged with an inert gas.
   C. the ports and charging lines should be capped.
   D. Both A and C are correct.

2. Gauge manifolds:
   A. have a small adjustment screw that allows the gauge to be calibrated.
   B. can never be calibrated.
   C. are oiled every week.
   D. Both A and C are correct.

3. Gauge manifolds should only be used with:
   A. refrigerant.
   B. clean oil.
   C. Both A and B are correct.
   D. None of the above is correct.

4. A compound gauge measures both pressure and vacuum:
   A. True.
   B. False.

RECOVERY UNIT

LABORATORY OBJECTIVE

The student will demonstrate how to properly use a refrigerant recovery unit.

LABORATORY NOTES

For this lab exercise there should be a typical recovery unit for the student to inspect.

FUNDAMENTALS OF HVAC/R TEXT REFERENCE

Unit 26

Required Tools and Equipment

Refrigerant recovery unit	Unit 26

SAFETY REQUIREMENTS

None

PROCEDURE

STEP 1. Locate a refrigerant recovery unit and examine it carefully so that you may complete the following exercise.

Refer to the *Fundamentals of HVAC/R* text Unit 26 for additional information.

    A. What types of refrigerants can the recovery unit be used for?

    _____

    _____

    B. Does the unit have a signal device to indicate when recovery has been completed?

    _____

    _____

C. Can the unit recover vapor refrigerant?

_____

_____

D. Draw a sketch representing a vapor recovery process in the space provided below.

E. If the recovery unit has a fan switch, what is it used for?

_____

_____

F. If you used this recovery unit to recover refrigerants, would this be considered system dependent (passive) or self contained (active)?

_____

_____

G. Could you perform liquid recovery with this unit?

_____

_____

H. Draw a sketch representing a vapor recovery process in the space provided below.

I. What is the primary difference between using a recovery unit as compared to a vacuum pump?

_____

_____

QUESTIONS

To get help in answering some of the following questions, refer to the *Fundamentals of HVAC/R* text Unit 26.

(Circle the letter that indicates the correct answer.)

1. A refrigerant recovery unit will stop on:
   A. low pressure.
   B. high pressure.
   C. Both A and B are correct.
   D. None of the above is correct.

2. Refrigerant recovery units:
   A. can only draw in vapor.
   B. can only draw in liquid.
   C. can draw in both liquids and vapors.
   D. All of the above are correct.

3. A liquid recovery is also known as the pull method:
   A. True.
   B. False.

VACUUM PUMP

LABORATORY OBJECTIVE

The student will demonstrate how to properly use a vacuum pump.

LABORATORY NOTES

For this lab exercise there should be a typical vacuum pump for the student to inspect.

FUNDAMENTALS OF HVAC/R TEXT REFERENCE

Unit 27

Required Tools and Equipment

Vacuum pump	Unit 27

SAFETY REQUIREMENTS

None

PROCEDURE

STEP 1. Locate a vacuum pump and examine it carefully so that you may complete the following exercise.

Refer to the *Fundamentals of HVAC/R* text Unit 27 for additional information.

    A. What is the rating of the vacuum pump in CFM?

_____

_____

B. Is the pump most suitable for residential work, small appliance work, or commercial work? How can you tell?

_____

_____

_____

C. Is the pump a deep vacuum pump?

_____

_____

D. Is it a single stage or a double stage pump?

_____

_____

E. Many vacuum pumps measure vacuum in microns. One inch of vacuum is how many microns?

_____

_____

F. Which is a deeper vacuum, 750 or 500 microns?

_____

_____

G. A deep vacuum pump should be at least how many microns?

_____

_____

QUESTIONS

To get help in answering some of the following questions, refer to the *Fundamentals of HVAC/R* text Unit 27.

(Circle the letter that indicates the correct answer.)

1. A vacuum pump is used to remove:
   A. noncondensable gases.
   B. refrigerant.
   C. oil.
   D. All of the above are correct.

2. Many vacuum pumps:
   A. use a water seal.
   B. have oil that needs to be periodically changed.
   C. are cooled by refrigerant.
   D. All of the above are correct.

3. Before evacuating a refrigeration system:
   A. fill it completely with an inert gas.
   B. run the vacuum pump for at least 1 hr to warm it up.

C. recover all of the refrigerant from the system.

D. None of the above is correct.

4. The deeper the vacuum pulled on a system:
   A. the wetter it will become.
   B. the drier it will become.
   C. the hotter it will become.
   D. the more brittle it will become.

5. At a low pressure:
   A. water will condense.
   B. oil will condense.
   C. water will vaporize.
   D. the vacuum pump will start automatically.

6. 20,080 microns is equal to about:
   A. 20,080 psia.
   B. 20,080 kPa.
   C. 40 psia.
   D. 0.4 psia.

7. Water can boil at:
   A. 212°F only, regardless of pressure.
   B. 100°C only, regardless of pressure.
   C. Both A and B are correct.
   D. at a temperature of −60°F at a pressure of 25 microns.

CHARGING CYLINDER

LABORATORY OBJECTIVE

The student will demonstrate how to properly use a refrigerant charging cylinder.

LABORATORY NOTES

For this lab exercise there should be a typical refrigerant charging cylinder for the student to inspect.

FUNDAMENTALS OF HVAC/R **TEXT REFERENCE**

Unit 27

Required Tools and Equipment

Charging cylinder	Unit 27

SAFETY REQUIREMENTS

None

PROCEDURE

STEP 1. Locate a charging cylinder and examine it carefully so that you may complete the following exercise.

Refer to the *Fundamentals of HVAC/R* text Unit 27 for additional information.

    A. What types of refrigerants can the cylinder be used for?

    _____

    _____

    B. Would you fill the charging cylinder with liquid or vapor refrigerant?

    _____

    _____

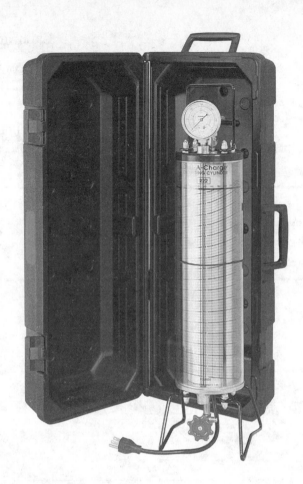

Figure F-7-1

C. What does the scale on the charging cylinder measure and what effect, if any, does room temperature have on its measurement?

_____

_____

_____

_____

D. Why would you electrically heat a charging cylinder?

_____

_____

QUESTIONS

To get help in answering some of the following questions, refer to the *Fundamentals of HVAC/R* text Unit 27.

(Circle the letter that indicates the correct answer.)

1. A refrigerant charging cylinder is calibrated:
   A. in units of weight.
   B. in units of temperature.

C. in units of volume.
D. in units of pressure.

2. Charging cylinders should always have a:
   A. heater.
   B. relief valve.
   C. sliding scale to adjust for temperature.
   D. All of the above are correct.

3. A charging cylinder can be used for any type of refrigerant.
   A. True.
   B. False.

4. The volume of liquid refrigerant in a charging cylinder will change with temperature.
   A. True.
   B. False.

5. A charging cylinder is usually filled:
   A. with vapor through the top.
   B. with liquid through the top.
   C. with vapor through the bottom.
   D. with liquid through the bottom.

6. Charging cylinder sizes generally range from:
   A. 8 to 10 oz of refrigerant.
   B. 16 to 32 oz of refrigerant.
   C. 2.5 to 10 lb of refrigerant.
   D. 1 to 15 lb of refrigerant.

ELECTRONIC LEAK DETECTOR

LABORATORY OBJECTIVE

The student will demonstrate how to properly use an electronic leak detector.

LABORATORY NOTES

For this lab exercise there should be a typical electronic leak detector for the student to inspect.

FUNDAMENTALS OF HVAC/R TEXT REFERENCE

Unit 27

Required Tools and Equipment

Electronic leak detector	Unit 27

SAFETY REQUIREMENTS

None

PROCEDURE

STEP 1. Locate an electronic leak detector and examine it carefully so that you may complete the following exercise.

Refer to the *Fundamentals of HVAC/R* text Unit 27 for additional information.

    A. What types of refrigerant can the leak detector be used for?

    _____

    _____

    B. Should the probe tip be positioned above or below the suspected leak?

    _____

    _____

Probe

Figure F-8-1

C. How quickly should the probe tip be moved across a suspected leak area?

_____

_____

D. How do you adjust the sensitivity of the meter?

_____

_____

E. Would the wind affect your reading outside?

_____

_____

QUESTIONS

To get help in answering some of the following questions, refer to the *Fundamentals of HVAC/R* text Unit 27.

(Circle the letter that indicates the correct answer.)

1. An electronic leak detector is capable of detecting a vacuum leak.
   A. True.
   B. False.

2. An electronic leak detector is more sensitive than a halide torch.
   A. True.
   B. False.

3. An electronic leak detector needs its sensitivity adjusted so that you can keep resetting it to get closer and closer to the source of the leak.
   A. True.
   B. False.

4. An electronic leak detector has a:
   A. copper button.
   B. titanium flapper.

C. platinum diode.
D. All of the above are correct.

5. If a system is pressurized with an inert gas:
   A. an electronic leak detector will find any leaks present.
   B. an electronic leak detector will not work.
   C. an electronic leak detector might work.
   D. None of the above is correct.

6. In order to use an electronic leak detector:
   A. the refrigeration system must be under a vacuum.
   B. the refrigeration system must be completely drained.
   C. the refrigeration system must contain CFCs.
   D. the refrigeration system must be under a positive pressure.

7. An electronic leak detector can detect very small refrigerant leaks.
   A. True.
   B. False.

8. When using an electronic leak detector to find a leak in a system that has been pressurized with an inert gas:
   A. a trace amount of R-22 must be added.
   B. a trace amount of argon must be added.
   C. a slight amount of paraffin based oil must be added.
   D. the system will need to be heated.

9. When raising the pressure on an air conditioning chiller to check for leaks:
   A. gag the pressure relief valve.
   B. warm the chill water loop.
   C. never exceed the rupture disk pressure.
   D. Both B and C are correct.

HALIDE TORCH LEAK DETECTOR

LABORATORY OBJECTIVE
The student will demonstrate how to properly use a halide torch leak detector.

LABORATORY NOTES
For this lab exercise there should be a typical halide torch leak detector for the student to inspect.

FUNDAMENTALS OF HVAC/R TEXT REFERENCE
Unit 27

Required Tools and Equipment

Halide torch leak detector	Unit 27

SAFETY REQUIREMENTS
None

PROCEDURE

STEP 1. Locate a halide torch leak detector and examine it carefully so that you may complete the following exercise. Refer to the *Fundamentals of HVAC/R* text Unit 27 for additional information.

    A. What color would you expect the flame to be when not exposed to refrigerants?

    _____

    _____

    B. What type of metal is exposed to the flame area?

    _____

    _____

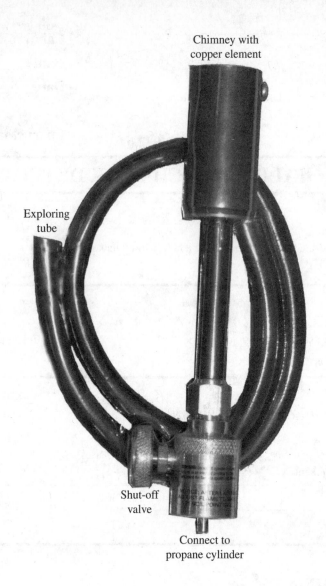

Figure F-9-1

C. What color would you expect the flame to be when exposed to a halide refrigerant?

_____

_____

D. According to EPA regulations, do systems with less than 50 lb of refrigerant have to be repaired?

_____

_____

E. Could the halide torch be used to detect ammonia refrigerant?

_____

_____

32

QUESTIONS

To get help in answering some of the following questions, refer to the *Fundamentals of HVAC/R* text Unit 27.

(Circle the letter that indicates the correct answer.)

1. A halide torch can be used for detecting:
    A. a small leak in a large area.
    B. a large leak in a small area.
    C. ammonia leaks.
    D. vacuum leaks.

2. A halide torch can be used for detecting:
    A. HFC refrigerants.
    B. ammonia.
    C. halocarbon refrigerants.
    D. inert gas leaks.

3. If the halide torch detects a refrigerant the flame color will change:
    A. green to blue.
    B. yellow to blue.
    C. blue to yellow.
    D. blue to green.

4. If the exploring tube of a halide torch is partially blocked:
    A. the flame will go out.
    B. the flame will turn white.
    C. the flame will smoke black.
    D. None of the above is correct.

5. The fuel for a halide torch can be:
    A. propane.
    B. butane.
    C. methyl alcohol.
    D. All of the above are correct.

6. When a halocarbon vapor passes over the hot copper element in a halide torch, the flame changes from normal color to bright green or purple.
    A. True.
    B. False.

SCHRADER, PIERCING, AND SERVICE VALVES

LABORATORY OBJECTIVE

The student will demonstrate how to properly replace the core on a refrigerant Schrader valve, install a piercing valve, and operate a service valve.

LABORATORY NOTES

For this lab exercise there should be a typical Schrader valve assembly connected to a low pressure air line as depicted below. There should also be a piercing valve and a service valve available for the student to use.

FUNDAMENTALS OF HVAC/R TEXT REFERENCE
Unit 25

Required Tools and Equipment

Refrigerant Schrader valve and tubing	Unit 25
Refrigerant piercing valve	Unit 25
Refrigerant service valve	Unit 25

SAFETY REQUIREMENTS
None

PROCEDURE

STEP 1. The inner core for a typical refrigerant Schrader valve will be replaced.

Refer to the *Fundamentals of HVAC/R* text Unit 25 for additional information.

A. Familiarize yourself with the operation of a Schrader valve. Notice the valve opens when the valve stem is depressed as shown in Figure F-10-1. When the stem is depressed, the system is open and at that point refrigerant could leak out of a system or if at a negative pressure air could leak in.

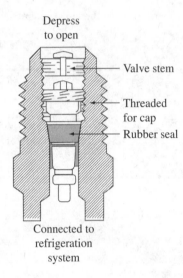

Depress to open

Valve stem

Threaded for cap

Rubber seal

Connected to refrigeration system

Figure F-10-1

B. The Schrader valve should be connected to a low pressure air supply with a shutoff valve as shown in Figure F-10-2. Close the air supply isolation valve and remove the core from the Schrader valve with the removal tool or if the valve cap is so designed, turn it around and use it to remove the core (unscrew counterclockwise).

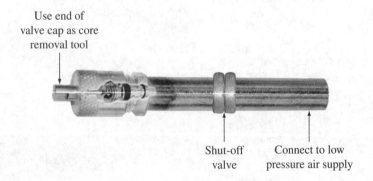

Use end of valve cap as core removal tool

Shut-off valve

Connect to low pressure air supply

Figure F-10-2

C. Install a replacement core and then test for leakage by turning the air supply back on to the assembly and apply a soap solution to the stem area and watch for bubbles forming.

Schrader valve cores

Figure F-10-3

D. Some core removal tools, as shown in Figure F-10-4, allow you to change the valve core while the system is under pressure. If you have such a tool, then repeat Parts B and C above without shutting off the air supply.

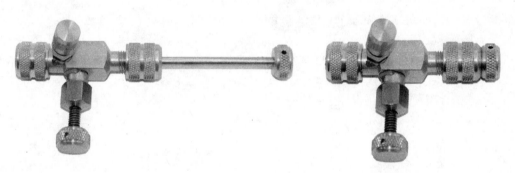

Figure F-10-4

STEP 2. Piercing valves: You will use the same tubing assembly from Step 1 and install a piercing valve. Check with your instructor regarding placement of the valve so as not to waste too much tubing.

Install piercing
valve wherever
most convenient

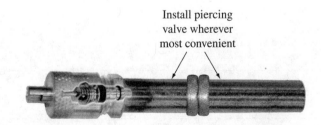

Figure F-10-5

A. Piercing valves are clamped to the tubing, sealed by a bushing gasket, and then they pierce the tube with a tapered needle. Some examples are shown in Figure F-10-6.

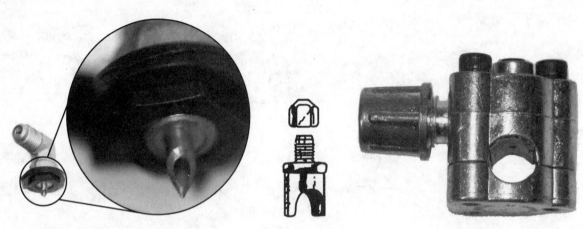

Figure F-10-6

B. Once the piercing valve is in place, it should be tested for leaks by applying a soap solution around the seal area and watching for bubbles.

STEP 3. Service valves: Locate a service valve and examine it carefully so that you may complete the following exercise. Refer to the *Fundamentals of HVAC/R* text Unit 25 for additional information.

A. Draw a cross sectional sketch of the service valve, labeling the components and ports in the space provided below.

B. Why is it called a service valve?

_____

_____

_____

_____

C. Why does the valve have a protective cap?

_____

_____

D. What type of packing does the valve have?

_____

_____

E. What is meant by back seating the valve and what is its purpose?

_____

_____

_____

_____

QUESTIONS

To get help in answering some of the following questions, refer to the *Fundamentals of HVAC/R* text Unit 25.

(Circle the letter that indicates the correct answer.)

1. When a service valve is back seated:
   A. the gauge manifold can be attached.
   B. normal flow through the valve will be shut off.
   C. there is less chance for leakage along the valve stem.
   D. Both A and C are correct.

2. A Schrader valve core can never be replaced while under pressure.
   A. True.
   B. False.

3. Access-piercing valves must be removed once the source of the sealed system malfunction has been located.
   A. True.
   B. False.

4. When a service valve is said to be in the mid-position:
   A. there is flow through the service port only.
   B. there is flow through the main valve only.
   C. there is flow through the service port and through the main valve.
   D. None of the above is correct.

5. The thing to always remember when conforming with EPA refrigerant management requirements is to:
   A. always release refrigerant to the atmosphere whenever possible.
   B. only vent refrigerant a little at a time.
   C. put service caps back on all access ports.
   D. None of the above is correct.

DESCRIBE THE REFRIGERATION SYSTEM CHARACTERISTICS USING A REFRIGERATION TRAINER

LABORATORY OBJECTIVE

The purpose of this lab is to demonstrate your ability to discuss the function of the four main refrigeration system components and describe the refrigerant characteristics throughout the system on the refrigeration trainer.

LABORATORY NOTES

You will be assigned a refrigeration trainer. You should be prepared to show the instructor each of the four main refrigeration components in order and discuss their function in the system. You will also discuss the characteristics of the refrigerant as it travels through the system using pressure and temperature measurements to help you determine the condition of the refrigerant throughout the trainer.

FUNDAMENTALS OF HVAC/R TEXT REFERENCE

Units 17 and 18

Required Tools and Equipment

Instrumented refrigerant trainer	
Pressure temperature chart	Appendix A
Temperature sensor with temperature clamp and thermocouple clamp	Unit 12

SAFETY REQUIREMENTS

None

PROCEDURE

STEP 1. Identify the components on the refrigeration trainer. Locate the compressor, condenser, metering device, and evaporator. Draw a sketch of the trainer. Label the components. Draw arrows to indicate the direction of refrigerant flow. Identify high pressure and low pressure circuits.

Figure F-11-1

DRAW SKETCH OF REFRIGERATION TRAINER IN THE SPACE PROVIDED

INSTRUCTOR CHECK BEFORE PROCEEDING TO NEXT STEP _____

STEP 2. Line up the components and then start the refrigeration trainer using the trainer's instruction manual if necessary. If you are not sure how to start the trainer, then check with your instructor.

STEP 3. Run the trainer and observe its operation. Note the pressure gauges that tell you what the refrigerant pressure is at every point in the system.

A. If there are no installed temperature sensors then you can measure the temperature at each location using a portable temperature sensor.

B. You may also touch components to get a feel for their temperature.

CAUTION: *Some components are HOT! Proceed carefully!*

C. If you hold your hand just above a warm component you can tell if it is extremely hot or not. If you can feel radiant heat coming from it, it may be too hot to touch safely.

STEP 4. Use the provided chart to record and organize your data. However, the job is not complete until you TELL your instructor how the system works and respond to questions.

    A. Record the pressure and temperature at each system location indicated on the chart.

    B. Compare the pressure and temperature using a pressure-temperature chart as found in Appendix A of the *Fundamentals of HVAC/R* text to determine the condition. These are saturated, superheated, or subcooled.

    C. Write in the state: gas, liquid, or mixed.

    D. Use the pressure, temperature, condition, and state to look up the heat content on an enthalpy chart such as found in Unit 18 of the *Fundamentals of HVAC/R* text.

Location	Temperature	Pressure	Condition	State	Heat Content
Compressor in					
Compressor out					
Discharge line					
Condenser in					
Condenser center					
Condenser out					
Liquid line					
Metering device in					
Metering device out					
Evaporator in					
Evaporator center					
Evaporator out					
Suction line					

STEP 5. Use the following terms to describe the changes in the refrigerant through each of the four major components and fill in the chart provided.

> *Stable:* no significant change in a condition
>
> *Small increase:* an increase of 30% or less
>
> *Large increase:* an increase of 200% or more
>
> *Small decrease:* a decrease of 30% or less
>
> *Large decrease:* a decrease of 200% or more

Type of Change	Compressor	Condenser	Metering Device	Evaporator
Change in pressure				
Change in temperature				
Change in heat				
Change in state				
Change in volume				

IDENTIFYING REFRIGERATION SYSTEM COMPONENTS

LABORATORY OBJECTIVE

The purpose of this lab is to demonstrate your ability to recognize each of the four main refrigeration system components on refrigeration systems.

LABORATORY NOTES

You will be assigned four typical refrigeration systems.

The units should be labeled SYSTEM 1, SYSTEM 2, SYSTEM 3, and SYSTEM 4.

You will locate the four major refrigeration system components in each system and complete a data sheet for each unit.

You will be prepared to show the instructor each of these components and discuss the basic characteristics of each of these components.

FUNDAMENTALS OF HVAC/R TEXT REFERENCE
Unit 17

Required Tools and Equipment

Packaged refrigeration unit	
Split system	
Small commercial refrigeration unit	
Any other unit of the instructor's choosing	
Six-in-one screwdriver for removing system panels	Unit 10

SAFETY REQUIREMENTS

Be careful of sharp edges when removing sheet metal panels.

PROCEDURE

STEP 1. Examine each unit and complete the appropriate data sheet. Remove any panels necessary to gain access to the unit components.

DATA SHEET SYSTEM 1

1. Point out the four main refrigeration components to your instructor.

   Instructor Check _____

2. What is the name and model number of this piece of equipment?

   _____

   _____

3. Is this a packaged unit or a split system?

   _____

4. This equipment would be installed where?

   _____

5. What style is the compressor: hermetic, semi-hermetic, or open?

   _____

6. Is the condenser air cooled or water cooled?

   _____

7. For air cooled condensers: Is the condenser induced draft or forced draft?

   _____

8. For water cooled condensers: Is this a waste water system or a recirculation type system?

   _____

9. Does the evaporator cool air or water?

   _____

10. What type of metering device does this system have: expansion valve, capillary tube, orifice, or other?

    _____

11. Does this system use a liquid receiver? Point it out.

    Instructor Check _____

12. Does this system have a suction accumulator? Point it out. What type of metering device does this system have: expansion valve, capillary tube, orifice, or other?

    Instructor Check _____

DATA SHEET SYSTEM 2

1. Point out the four main refrigeration components to your instructor.

   Instructor Check _____

2. What is the name and model number of this piece of equipment?

   _____

   _____

3. Is this a packaged unit or a split system?

   _____

4. This equipment would be installed where?

   _____

5. What style is the compressor: hermetic, semi-hermetic, or open?

   _____

6. Is the condenser air cooled or water cooled?

   _____

7. For air cooled condensers: Is the condenser induced draft or forced draft?

   _____

8. For water cooled condensers: Is this a waste water system or a recirculation type system?

   _____

9. Does the evaporator cool air or water?

   _____

10. What type of metering device does this system have: expansion valve, capillary tube, orifice, or other?

    _____

11. Does this system use a liquid receiver? Point it out.

    Instructor Check _____

12. Does this system have a suction accumulator? Point it out. What type of metering device does this system have: expansion valve, capillary tube, orifice, or other?

    Instructor Check _____

DATA SHEET SYSTEM 3

1. Point out the four main refrigeration components to your instructor.

   Instructor Check _____

2. What is the name and model number of this piece of equipment?

   _____

   _____

3. Is this a packaged unit or a split system?

   _____

4. This equipment would be installed where?

   _____

5. What style is the compressor: hermetic, semi-hermetic, or open?

   _____

6. Is the condenser air cooled or water cooled?

   _____

7. For air cooled condensers: Is the condenser induced draft or forced draft?

   _____

8. For water cooled condensers: Is this a waste water system or a recirculation type system?

   _____

9. Does the evaporator cool air or water?

   _____

10. What type of metering device does this system have: expansion valve, capillary tube, orifice, or other?

    _____

11. Does this system use a liquid receiver? Point it out.

    Instructor Check _____

12. Does this system have a suction accumulator? Point it out. What type of metering device does this system have: expansion valve, capillary tube, orifice, or other?

    Instructor Check _____

DATA SHEET SYSTEM 4

1. Point out the four main refrigeration components to your instructor.

   Instructor Check _____

2. What is the name and model number of this piece of equipment?

   _____

   _____

3. Is this a packaged unit or a split system?

   _____

4. This equipment would be installed where?

   _____

5. What style is the compressor: hermetic, semi-hermetic, or open?

   _____

6. Is the condenser air cooled or water cooled?

   _____

7. For air cooled condensers: Is the condenser induced draft or forced draft?

   _____

8. For water cooled condensers: Is this a waste water system or a recirculation type system?

   _____

9. Does the evaporator cool air or water?

   _____

10. What type of metering device does this system have: expansion valve, capillary tube, orifice, or other?

    _____

11. Does this system use a liquid receiver? Point it out.

    Instructor Check _____

12. Does this system have a suction accumulator? Point it out. What type of metering device does this system have: expansion valve, capillary tube, orifice, or other?

    Instructor Check _____

IDENTIFYING REFRIGERANTS AND REFRIGERANT CHARACTERISTICS

LABORATORY OBJECTIVE

The purpose of this lab is to demonstrate your ability to identify the type of refrigerant contained in each of four different refrigeration systems.

LABORATORY NOTES

You will find the type of refrigerant, the amount of refrigerant, and the system test pressures on the unit data plate of each system. Then you will research the reference material to find the refrigerant characteristics.

You will identify the refrigerant by

- Name
- Pressure (very high, high, or low)
- Ozone depletion potential
- Global warming potential
- Toxicity
- Flammability
- Chemical composition
- Formulation

FUNDAMENTALS OF HVAC/R TEXT REFERENCE
Unit 23

Required Tools and Equipment

Four refrigeration systems containing different refrigerants	
Refrigerant cylinders	Unit 26
Refrigerant MSDS	Unit 3

SAFETY REQUIREMENTS:

Caution!!! Do not open the valves on the refrigeration cylinders to avoid possible injury due to skin contact with refrigerant. Also it is illegal to intentionally vent refrigerants composed of CFCs and HCFCs.

PROCEDURE

STEP 1. Locate the name plate on the first unit. The name plate will be in different locations for each unit you inspect.

A. Find and record the unit model number on the provided Refrigerant Characteristics Chart for the first refrigeration system.

B. Find and record the refrigerant type on the provided Refrigerant Characteristics Chart for the first refrigeration system.

C. Find and record the high side test pressure on the provided Refrigerant Characteristics Chart for the first refrigeration system.

D. Find and record the low side test pressure on the provided Refrigerant Characteristics Chart for the first refrigeration system.

E. Locate a cylinder of the same refrigerant type and read the instructions on the cylinder.

F. Locate the MSDS for this refrigerant type and read the information regarding the refrigerant properties.

G. Use the information from the refrigerant cylinder, MSDS, *Fundamentals of HVAC/R* text Unit 23, and *Fundamentals of HVAC/R* text Appendix A to complete filling out the provided Refrigerant Characteristics Chart for the first refrigeration system.

STEP 2. Repeat the process from Step 1 for the other three refrigeration systems completing the provided Refrigerant Characteristics Chart for each one. When finished, have your instructor check the information that you recorded and then answer the questions at the end of this laboratory exercise.

QUESTIONS

1. Were there any refrigerants that did not have either an ozone depletion potential or a global warming potential?

   _____

   _____

2. What are the high and low side test pressures used for?

   _____

   _____

3. Which refrigerants may safely leave the cylinder as either a gas or a liquid?

   _____

   _____

Refrigerant Characteristics Chart

	System #1	System #2	System #3	System #4
Model number				
Refrigerant name				
Refrigerant quantity				
High side test pressure				
Low side test pressure				
Pressure (very high, high, low)				
Ozone depletion potential				
Global warming potential				
Safety rating				
Chemical composition				
Formulation				

4. Which refrigerants must leave the cylinder as a liquid only?

_____

_____

5. Which refrigerant had the highest system test pressures?

_____

_____

6. Which refrigerant had the lowest system test pressures?

_____

_____

7. What hazards do you need to be aware of when working with these refrigerants?

_____

_____

_____

_____

_____

_____

_____

8. Which of these refrigerants will NOT be used in equipment of the future?

_____

_____

9. Which of these refrigerants WILL be used in equipment of the future?

_____

_____

IDENTIFYING REFRIGERANT LUBRICANT CHARACTERISTICS

LABORATORY OBJECTIVE

The purpose of this lab is to demonstrate your ability to identify three different types of refrigerant lubricant and to recommend the proper lubricant for four refrigeration systems.

LABORATORY NOTES

You will examine three different types of refrigeration lubricant and recommend where they can be used. You will recommend the correct type of lubricant for four refrigeration systems.

FUNDAMENTALS OF HVAC/R TEXT REFERENCE

Unit 23

Required Tools and Equipment

Four refrigeration systems containing different refrigerants	
Three containers of different refrigeration lubricant	

SAFETY REQUIREMENTS

Caution!!! Do not open the lubricant containers and spill lubricant or allow it to contact your skin.

PROCEDURE

STEP 1. Locate the first container of refrigerant lubricant.

    A. Identify the type of lubricant in the first container.

    B. Record in the Lubricant Type Chart provided:

        • Type of lubricant
        • Lubricant viscosity

- Recommended evaporator temperature range
- Type of refrigerants that the lubricant is compatible with

C. Repeat this process for the other lubricant types.

Lubricant Type Chart

	Brand Name	Type (Mineral, PAG, POE, AB)	Viscosity	Evaporator Temperature	Refrigerant Compatibility
Container #1					
Container #2					
Container #3					

STEP 2. Locate the name plate on the first refrigeration unit. The name plate will be in different locations for each unit you inspect.

   A. Find and record the unit model number on the provided System Refrigerant Lubricant Characteristics Chart for the first refrigeration system.

   B. Find and record the refrigerant type on the provided System Refrigerant Lubricant Characteristics Chart for the first refrigeration system.

   C. Recommend a lubricant that is compatible with this unit.

STEP 3. Repeat the process from Step 2 for the other three refrigeration systems completing the provided System Refrigerant Lubricant Characteristics Chart for each one. When finished have your instructor check the information that you recorded and then answer the questions at the end of this laboratory exercise.

System Refrigerant Lubricant Characteristics Chart

	System #1	System #2	System #3	System #4
Model number				
Refrigerant name				
Recommended lubricant				

QUESTIONS

   1. Which lubricant can be used with the widest range of systems?

   _____

   _____

2. What is the difference in viscosity between lubricants designed for higher evaporator temperatures and lubricants that are designed for use in lower temperature systems?

_____

_____

_____

_____

_____

3. What is the most common viscosity for lubricants used in air conditioning systems?

_____

_____

_____

QUANTITY-PRESSURE RELATIONSHIP OF IDEAL GASES

LABORATORY OBJECTIVE

The student will demonstrate that the pressure of an ideal gas contained in a cylinder depends upon the amount of gas in the cylinder.

LABORATORY NOTES

We will compare the pressure of a cylinder with two different weights of the same gas at the same temperature. Using the same cylinder will ensure that the volume remains constant. The temperature will be verified using an infrared thermometer.

FUNDAMENTALS OF HVAC/R TEXT REFERENCE

Unit 8

Required Tools and Equipment

An empty refillable refrigerant recovery cylinder	Unit 26
Vacuum pump	Unit 27
Digital scale	Unit 26
Cylinder of nitrogen gas with regulator	
Infrared thermometer	Unit 12
Gauge manifold	Units 12 and 25

SAFETY REQUIREMENTS

A. Wear safety glasses and gloves whenever handling gas cylinders and regulators.

B. Familiarize yourself with the proper procedures for operating gas cylinder regulators.

PROCEDURE

STEP 1. Locate an empty refillable refrigerant recovery cylinder, a nitrogen cylinder with a regulator, a vacuum pump, a gauge manifold, and a digital refrigerant charging scale. Refer to Laboratory Worksheets F-3 and F-6 if necessary.

A. Place the refrigerant recovery cylinder on the digital scale and connect the vacuum pump using the gauge manifold. Also connect the gauge manifold to the nitrogen cylinder but keep the nitrogen cylinder valves closed as shown in Figure P-1-1.

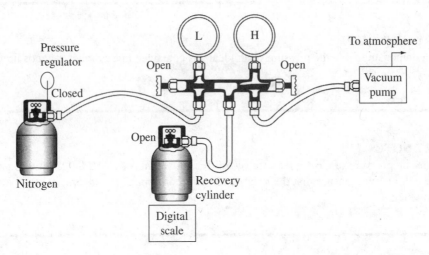

Figure P-1-1

B. Run the vacuum pump to evacuate the cylinder and gauge manifold hoses to ensure that the cylinder and hoses are empty and free from all gas. Close the high side valve (H) on the gauge manifold before shutting off the vacuum pump or else air will leak back into the lines and cylinder.

C. Close the low side valve (L) on the gauge manifold.

D. Zero the digital scale.

E. The valve arrangement should now be as shown in Figure P-1-2.

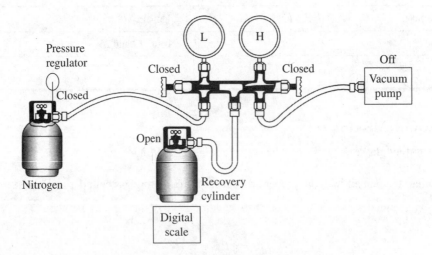

Figure P-1-2

62

STEP 2. Open the nitrogen cylinder and adjust the pressure regulator as follows:

A. Make sure that the nitrogen cylinder pressure regulator is turned all the way out (counterclockwise).

B. Slowly open the cylinder valve fully open to back seat it. The tank pressure should register on the regulator high pressure gauge. The pressure in the tank can be in excess of 2,000 psi. **Caution!** Do not stand in front of the regulator T handle.

C. Slowly turn the regulator T handle inward (clockwise) until the regulator adjusted pressure reaches approximately 50 psig.

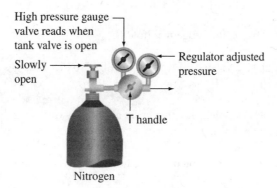

Figure P-1-3

STEP 3. Slowly open the low side valve on the gauge manifold (L) until you see a weight increase on the digital scale. Then close the gauge manifold valve and record your readings.

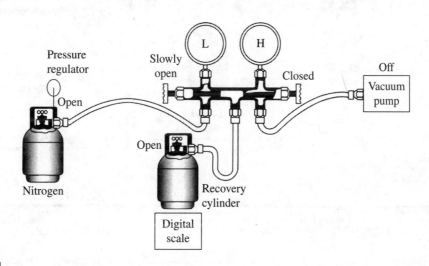

Figure P-1-4

Gas pressure _____

Gas temperature _____

Gas weight _____

STEP 4. Once again slowly open the low side valve on the gauge manifold (L) and add more nitrogen until you see a weight increase on the digital scale. Then close the gauge manifold valve and record your readings.

Gas pressure _____

Gas temperature _____

Gas weight _____

STEP 5. After recording the second set of readings you can prepare to disconnect the gauge manifold as follows:

A. Close the shut-off valve on the nitrogen cylinder.

B. Disconnect the vacuum pump from the gauge manifold.

C. Open the low side valve on the gauge manifold (L).

D. Slowly open the high side valve on the gauge manifold to bleed the gas pressure from the recovery cylinder and all hoses.

E. Once the pressure has been bled, you may back all the way off on the T handle (counterclockwise) for the pressure regulator on the nitrogen cylinder.

F. After all of the pressure has been bled off, the hoses may be disconnected.

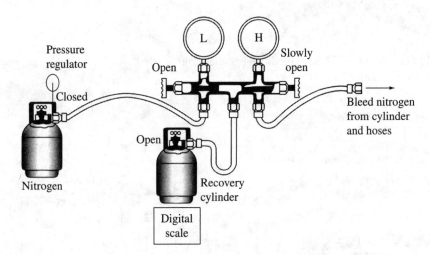

Figure P-1-5

QUESTIONS

1. How does the pressure of the cylinder change when the amount of nitrogen changes?

_____

_____

_____

_____

2. What conclusions can you draw about the behavior of an ideal gas?

_____

_____

_____

_____

GAS TEMPERATURE AND VOLUME AT CONSTANT PRESSURE

LABORATORY OBJECTIVE
The purpose of this lab is to demonstrate the effect of temperature change on the volume of a gas that is at constant pressure.

LABORATORY NOTES
The gas will be contained in a mylar balloon. The balloon's volume can change because the balloon can expand and contract. The pressure inside the balloon will stay the same as will the atmospheric pressure surrounding the balloon. We will change the temperature of the gas and observe the volume change by observing the size of the balloon.

FUNDAMENTALS OF HVAC/R TEXT REFERENCE
Unit 8

Required Tools and Equipment

Mylar balloon	
Freezer	
Hair dryer	

SAFETY REQUIREMENTS

A. Wear safety glasses and gloves whenever handling gas cylinders and regulators.

B. Familiarize yourself with the proper procedures for operating gas cylinder regulators.

PROCEDURE

STEP 1. Inflate the mylar balloon at room temperature until it is full, but not taut.

STEP 2. Place the balloon in the freezer for 5 min.

STEP 3. Remove the balloon and compare its shape and size to its original shape and size.

STEP 4. Heat the balloon with a hair dryer.

STEP 5. Compare the balloon's shape and size to its original shape and size.

QUESTIONS

1. What happened to the balloon when the temperature dropped?

   _____

   _____

   _____

   _____

2. What happened to the balloon when the temperature rose?

   _____

   _____

   _____

   _____

3. What does this tell us about the volume of a gas compared to its temperature?

   _____

   _____

   _____

   _____

GAS PRESSURE-TEMPERATURE RELATIONSHIP AT A CONSTANT VOLUME

LABORATORY OBJECTIVE

The purpose of this lab is to demonstrate the effect of gas temperature changes on gas pressure when the volume remains constant.

LABORATORY NOTES

The gas will be contained in a recovery cylinder. Since the cylinder is a fixed volume, the gas volume will not change. We will observe the pressure and temperature of the cylinder filled with nitrogen at room temperature. Then we will cool the cylinder in ice and observe the effect of reduced temperature on the gas pressure. Next, we will heat the cylinder and observe the effect of increased temperature on gas pressure. We will use the ideal gas law to verify our results.

FUNDAMENTALS OF HVAC/R TEXT REFERENCE
Unit 8

Required Tools and Equipment

Refrigerant recovery cylinder containing nitrogen	Unit 26
Mop bucket	
Ice	
Infrared thermometer	Unit 12
Gauge manifold	Units 12 and 25

SAFETY REQUIREMENTS

Wear safety glasses and gloves whenever handling gas cylinders and regulators.

PROCEDURE

STEP 1. Obtain a refrigerant recovery cylinder containing nitrogen gas.

A. Measure and record the pressure and temperature of the cylinder.

Pressure _____

Temperature _____

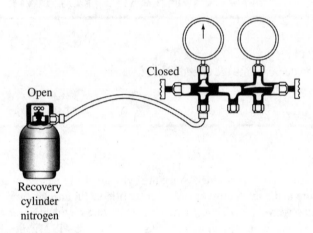

Figure P-3-1

STEP 2. Place the recovery cylinder into the mop bucket and pack it with ice.

A. Wait 15 min and then record the pressure and temperature of the cylinder.

Pressure _____

Temperature _____

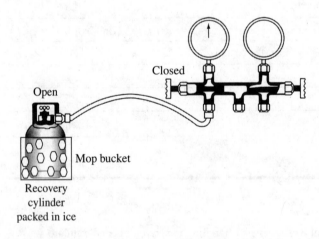

Figure P-3-2

STEP 3. Calculate the expected pressure after cooling the cylinder. For this calculation use the starting pressure, starting temperature, and ending temperature. Refer to the ideal gas laws from Unit 8 in the *Fundamentals of HVAC/R* text.

(Remember that all pressures and temperatures must be converted to absolute readings BEFORE working the problem.)

Calculated Pressure _____

Measured Pressure _____

A. What happened to the pressure in the cylinder when the temperature dropped?

_____

_____

_____

B. How do your measured results compare to your calculated results?

_____

_____

_____

STEP 4. Place the recovery cylinder into a sink and run warm water over the cylinder.

Caution!!! The refrigerant cylinder temperature should NEVER exceed 125°F.

A. Record the new pressure and temperature of the cylinder.

Pressure _____

Temperature _____

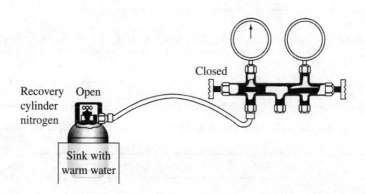

Figure P-3-3

STEP 5. Calculate the expected pressure after warming the cylinder. For this calculation use the starting pressure, starting temperature, and ending temperature. Refer to the ideal gas laws from Unit 8 in the *Fundamentals of HVAC/R* text.

(Remember that all pressures and temperatures must be converted to absolute readings BEFORE working the problem.)

SHOW ALL OF YOUR CALCULATIONS IN THE SPACE PROVIDED BELOW.

Calculated pressure _____

Measured pressure _____

A. What happened to the pressure in the cylinder when the temperature increased?

_____

_____

_____

B. How do your measured results compare to your calculated results?

_____

_____

_____

C. Based upon the results from cooling and then heating the cylinder, what conclusions can you draw about the pressure of a gas compared to its temperature at constant volume?

_____

_____

_____

_____

_____

_____

GAS PRESSURE-TEMPERATURE-VOLUME RELATIONSHIP

LABORATORY OBJECTIVE

The purpose of this lab is to demonstrate the combined effect of gas volume and pressure changes on gas temperature.

LABORATORY NOTES

First, we will observe the temperature change in a gas that is being compressed. During compression the volume will be reduced and the pressure will be increased. Second, we will observe the temperature change in a gas as it is expanded. During expansion the volume will increase and the pressure will decrease.

FUNDAMENTALS OF HVAC/R TEXT REFERENCE
Unit 8

Required Tools and Equipment

Empty refrigerant recovery cylinder	Unit 26
Bicycle pump	
Temperature sensor with thermocouple probe	Unit 12
Infrared thermometer	Unit 12
Gauge manifold	Units 12 and 25

SAFETY REQUIREMENTS

Wear safety glasses and gloves whenever handling gas cylinders and regulators.

PROCEDURE

STEP 1. Obtain an empty refrigerant recovery cylinder.

A. Measure and record the pressure and temperature of the cylinder. (The pressure should be atmospheric.)

Pressure _____

Temperature _____

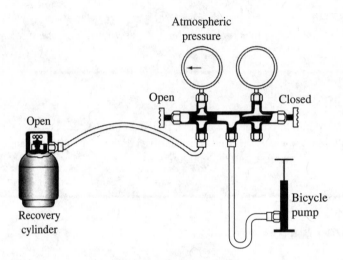

Figure P-4-1

STEP 2. Use the bicycle pump to raise the pressure of the cylinder to 40 psig, then close the valve on the cylinder and record the pressure and temperature of the cylinder.

Pressure _____

Temperature _____

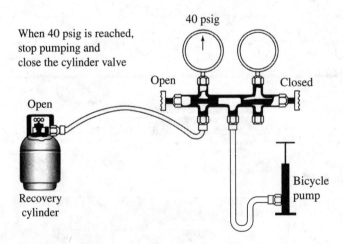

Figure P-4-2

Step 3. Wait for the cylinder to cool to room temperature. Remove the gauge manifold and then open the gas valve on the cylinder and measure the temperature of the air leaving the cylinder (the air is expanding).

Temperature _____

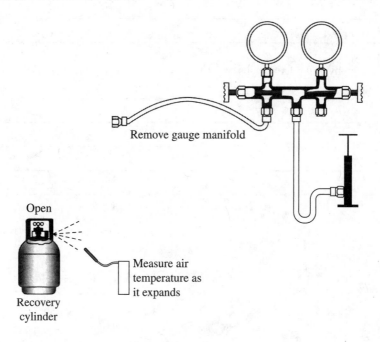

Remove gauge manifold

Open

Measure air temperature as it expands

Recovery cylinder

Figure P-4-3

QUESTIONS

1. What happened to the temperature of the air when it was compressed with the bicycle pump?

_____
_____
_____
_____

2. Why did this happen?

_____
_____
_____
_____

3. What happened to the temperature of the air when it was expanded?

_____
_____
_____
_____

4. Why did this happen?

_____

_____

_____

_____

5. How could these physical relationships of pressure-volume-temperature be useful in an air conditioning system?

_____

_____

_____

_____

_____

BOILING POINT VERSUS PRESSURE

LABORATORY OBJECTIVE

The purpose of this lab is to demonstrate the effect of pressure on the boiling point of a liquid.

LABORATORY NOTES

We know that the boiling point of water at atmospheric pressure is 212°F. We will use a vacuum pump to reduce the pressure on a flask of water and observe the effect of reduced pressure. A vacuum pump reduces the pressure of a closed container by removing gas from the container.

FUNDAMENTALS OF HVAC/R TEXT REFERENCE
Unit 7

Required Tools and Equipment

Flask with rubber stopper	
Vacuum gauge that reads in inches of mercury vacuum (in Hg)	Unit 27
Vacuum pump	Unit 27

SAFETY REQUIREMENTS
None.

PROCEDURE

STEP 1. Obtain an empty flask and rubber stopper.

    A. Fill the flask half full with water.

    B. Stopper the top of the flask.

    C. Connect the flask to the vacuum pump.

    D. Connect the vacuum gauge to the vacuum pump.

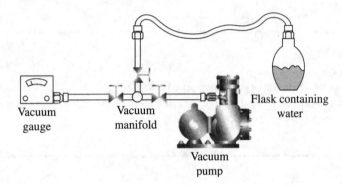

Figure P-5-1

STEP 2. Operate the vacuum pump and then answer the following questions.

    1. What happens to the pressure in the flask?

    _____

    _____

    _____

    _____

    2. What happens to the water in the flask?

    _____

    _____

    _____

    _____

    3. Explain why this happens.

    _____

    _____

    _____

    _____

STEP 3. Turn off the vacuum pump and then answer the following questions.

4. What happens to the pressure in the flask?

   _____

   _____

   _____

   _____

5. Why does this happen?

   _____

   _____

   _____

   _____

6. What happens to the water?

   _____

   _____

   _____

   _____

7. Why does this happen?

   _____

   _____

   _____

   _____

PRESSURE-TEMPERATURE RELATIONSHIP
OF SATURATED MIXTURES

LABORATORY OBJECTIVE

You will demonstrate the relationship of pressure and temperature for saturated gas-liquid mixtures.

LABORATORY NOTES

You will observe the behavior of the saturated gas by comparing the pressure of two cylinders containing different weights of a saturated mix at the same temperature. Then we will change the temperature and observe the effect on pressure.

You will be using two recovery cylinders with refrigerant. Both cylinders should be the same type of refrigerant, but they should contain different amounts of refrigerant. Both cylinders should be the same temperature and both *should contain some liquid and some vapor.*

FUNDAMENTALS OF HVAC/R TEXT REFERENCE

Unit 7

Required Tools and Equipment

Two refrigerant recovery cylinders with the same refrigerant type	Unit 26
Freezer or a bucket of ice	
Heat gun or a sink with hot water	
Gauge manifold	Unit 12
Infrared thermometer	Unit 12
Digital scale	Unit 27

SAFETY REQUIREMENTS

Wear safety goggles and gloves when working with refrigerants. Liquid refrigerant can cause frostbite when in contact with eyes or skin.

PROCEDURE

Step 1. Measure the temperature, pressure, and weight of the first refrigerant recovery cylinder as follows.

A. Place one of the refrigerant recovery cylinders on top of a digital scale and then connect the refrigerant recovery cylinder to the high side of the gauge manifold as shown in Figure P-6-1.

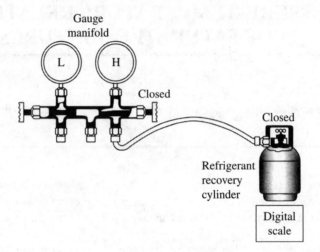

Figure P-6-1

B. Slowly open the refrigerant recovery cylinder valve as shown in Figure P-6-2.

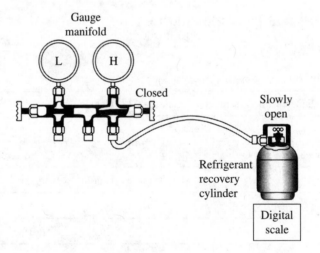

Figure P-6-2

C. Record the pressure from the reading on the high side gauge.

Recovery cylinder #1 pressure _____

D. Measure the cylinder temperature using the infrared thermometer.

Recovery cylinder #1 temperature _____

82

E. Record the cylinder weight from the reading on the digital scale.

Recovery cylinder #1 weight _____

F. After recording the cylinder pressure and temperature, close the cylinder valve as shown in Figure P-6-3.

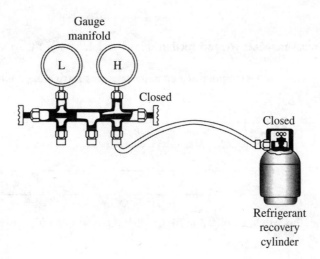

Figure P-6-3

G. You must now bleed off the pressure remaining in the hose before you disconnect it. To do this, make sure that the cylinder valve is closed. Next, take the center hose of the gauge manifold and direct it away from any person including yourself. Slowly open the high side gauge manifold valve as shown in Figure P-6-4 and the pressure in the line will bleed off. Be careful as the hose may whip around slightly as it drains.

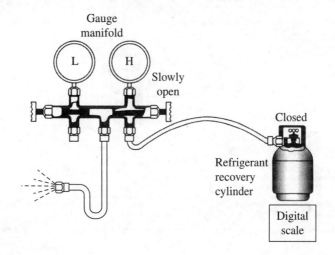

Figure P-6-4

STEP 2. Repeat Step 1 using the second refrigerant recovery cylinder.

A. Record the pressure from the reading on the high side gauge.

Recovery cylinder #2 pressure

B. Measure the cylinder temperature using the infrared thermometer.

Recovery cylinder #2 temperature _____

C. Record the cylinder weight from the reading on the digital scale.

Recovery cylinder #2 weight _____

STEP 3. Referring to the measurements you recorded in steps 1 and 2, answer the following questions.

1. How does the pressure of the cylinder that contains the most refrigerant compare to the cylinder with the least refrigerant?

_____

_____

_____

_____

2. Do you think that this would be different for a cylinder containing only gas without any liquid? Explain your answer.

_____

_____

_____

_____

STEP 4. Place one of the cylinders in a bucket of ice or in the freezer.

Recovery
cylinder
packed in ice

Figure P-6-5

A. Allow the cylinder to cool for 30 min and then measure its pressure and temperature as you did in Step 1.

B. Record the pressure from the reading on the high side gauge.

Cooled recovery cylinder pressure _____

C. Measure the cylinder temperature using the infrared thermometer.

Cooled recovery cylinder temperature _____

STEP 5. Calculate the expected pressure after cooling the cylinder. For this calculation use the starting pressure, starting temperature, and ending temperature. Refer to the ideal gas laws from Unit 8 in the *Fundamentals of HVAC/R* text.

(Remember that all pressures and temperatures must be converted to absolute readings BEFORE working the problem.)

3. How does the actual pressure compare to the pressure predicted by the gas law?

_____

_____

_____

_____

4. Use a saturation chart such as the one found in Appendix A from the *Fundamentals of HVAC/R* text and determine the pressure that corresponds to the temperature of the refrigerant in the cylinder.

Saturated pressure from chart _____

5. How does the pressure found from the chart compare to the actual pressure?

_____

_____

_____

_____

6. Which was more accurate in predicting the results, the ideal gas law or using the saturation chart?

_____

_____

_____

_____

STEP 6. Place the recovery cylinder into a sink and run warm water over the cylinder.

Caution!!! The refrigerant cylinder temperature should NEVER exceed 125°F.

Figure P-6-5

A. Allow the cylinder to warm for 15 min and then measure its pressure and temperature as you did in STEP 1.

B. Record the pressure from the reading on the high side gauge.

   Heated recovery cylinder pressure _____

C. Measure the cylinder temperature using the infrared thermometer.

   Heated recovery cylinder temperature _____

STEP 7. Calculate the expected pressure after cooling the cylinder. For this calculation use the starting pressure, starting temperature, and ending temperature. Refer to the ideal gas laws from Unit 8 in the *Fundamentals of HVAC/R* text.

(Remember that all pressures and temperatures must be converted to absolute readings BEFORE working the problem.)

SHOW ALL OF YOUR CALCULATIONS IN THE SPACE PROVIDED BELOW.

7. How does the actual pressure compare to the pressure predicted by the ideal gas law?

_____

_____

_____

_____

8. Use a saturation chart such as the one found in Appendix A from the _Fundamentals of HVAC/R_ text and determine the pressure that corresponds to the temperature of the refrigerant in the cylinder.

Saturated pressure from chart _____

9. How does the pressure found from the chart compare to the actual pressure?

_____

_____

_____

_____

10. Which was more accurate in predicting the results, the ideal gas law or using the saturation chart?

_____

_____

_____

_____

11. What conclusions can you draw about the behavior of saturated liquid-gas mixtures to temperature changes?

_____

_____

_____

_____

SENSIBLE AND LATENT HEAT

LABORATORY OBJECTIVE

You will demonstrate sensible and latent heat properties.

LABORATORY NOTES

We will demonstrate sensible and latent heat properties by adding heat to water while we monitor the amount of energy used, the temperature of the water, and the weight of the water. Energy input will be determined by measuring the wattage used, temperature change will be measured with a thermocouple, and weight change will be measured with a digital scale.

FUNDAMENTALS OF HVAC/R TEXT REFERENCE

Unit 7

Required Tools and Equipment

Temperature sensor with temperature clamp and thermocouple clamp	Unit 12
Electric cook pot	
Watt meter	Unit 11

SAFETY REQUIREMENTS

A. Wear safety goggles and gloves when working with high temperature liquids.

B. The steam that is generated can easily burn your skin. Be careful not to come in contact with the hot steam.

PROCEDURE

STEP 1. Obtain an electric cook pot and a digital scale.

A. Place the cook pot on top of the digital scale and pour approximately 3 lb of water into the pot.

B. Zero the digital scale with the pot and the water.

C. Measure the temperature of the water.

Water temperature _____

D. Connect the wattmeter to the cook pot circuit to measure the power that will be used during the heating of the water.

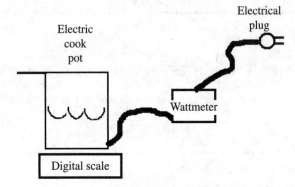

Figure P-7-1

E. Plug in the cook pot and record the time and wattage.

Time started

Wattage

STEP 2. Monitor the time, temperature, and wattage as follows.

A. When the temperature reaches approximately 150°F, record the time, the exact temperature, and the wattage.

Time

Temperature

Wattage

B. When water begins to boil, record the time, the exact temperature, and the wattage.

Time

Temperature

Wattage

C. When the water reaches a rolling boil, record the time, the exact temperature, the weight on the scale, and the wattage. Let the water boil for 3 min and then record these values again in part D.

Time

Temperature

Weight _____

Wattage _____

D. Record the values after 3 min of boiling.

Time _____

Temperature _____

Weight _____

Wattage _____

STEP 3. Calculate the amount of energy in Btu required to heat the water to 150°F.

A. Since it takes 1 Btu to heat 1 lb of water 1°F at atmospheric pressure, you can use the formula:

(weight of the water in pounds) × (1 Btu/lb-°F) × (ending temperature − beginning temperature) = Btu required to heat the water.

SHOW ALL OF YOUR CALCULATIONS IN THE SPACE PROVIDED BELOW.

TOTAL AMOUNT OF HEAT REQUIRED (Btu) _____

B. Now calculate the measured amount of electrical heat input. One watt is equivalent to 3.41 Btu per hour or divide this by 60 min in 1 hr for 0.057 Btu per minute. With this we can convert our measured wattage into Btu using the formula:

(measured wattage) $\times$ (minutes to 150°F) $\times$ (0.057 Btu/min-watt) = Btu input

SHOW ALL OF YOUR CALCULATIONS IN THE SPACE PROVIDED BELOW.

TOTAL AMOUNT OF HEAT INPUT (Btu) _____

1. How does the amount of heat required to heat the water in Btu compare to the measured amount of heat input? If they are not equal, then explain why.

_____

_____

_____

_____

STEP 4. Calculate the amount of energy in Btu required to boil the water for 3 min.

A. Since it takes 970 Btu to boil 1 lb of water at atmospheric pressure, you can use the formula:

(pounds of water beginning − pounds of water ending for the 3 min period) × (970 Btu/lb) = Btu required to boil the water for 3 min.

SHOW ALL OF YOUR CALCULATIONS IN THE SPACE PROVIDED BELOW.

TOTAL AMOUNT OF HEAT REQUIRED (Btu)

B. Now calculate the measured amount of electrical heat input. One watt is equivalent to 3.41 Btu per hour or divide this by 60 min in 1 hr for 0.057 Btu per minute. With this we can convert our measured wattage into Btu using the formula:

(measured wattage) × (3 min) × (0.057 Btu/min-watt) = Btu input

SHOW ALL OF YOUR CALCULATIONS IN THE SPACE PROVIDED BELOW.

TOTAL AMOUNT OF HEAT INPUT (Btu) _____

2. How does the amount of heat required to boil the water in Btu compare to the measured amount of heat input? If they are not equal, then explain why.

_____

_____

_____

_____

3. What happened to the temperature of the water as it was boiling?

_____

_____

4. What do you think the temperature of the steam would be?

_____

_____

5. Does it take more energy to heat the water or to boil the water?

_____

_____

DETERMINING REFRIGERANT CONDITION

LABORATORY OBJECTIVE

The purpose of this lab is to demonstrate your ability to determine if refrigerant is saturated, superheated, or subcooled.

LABORATORY NOTES

You will measure the pressure and temperature of several refrigeration system components. You will determine if the refrigerant in the component is saturated, superheated, or subcooled by comparing the pressure and temperature to a saturation chart for that particular refrigerant.

***FUNDAMENTALS OF HVAC/R* TEXT REFERENCE**

Units 7, 18, and 23

Required Tools and Equipment

Instrumented refrigerant trainer	
Pressure temperature chart	Appendix A

SAFETY REQUIREMENTS

None

PROCEDURE

STEP 1. Identify the components on the refrigeration trainer (refer to Lab F-11 *Describe the Refrigeration System Characteristics on a Refrigeration Trainer*).

STEP 2. Line up the components and then start the refrigeration trainer using the trainer's instruction manual if necessary.

STEP 3. Allow the refrigerant trainer to run until the temperatures and pressures have begun to stabilize and then complete the chart below. You will need to identify for each situation whether the refrigerant is subcooled, saturated, or superheated. Not all refrigeration trainers are instrumented the same, so if there is no reading available on the refrigerant trainer for the specific condition, then leave that space blank.

Location	Pressure	Temperature from Chart	Actual Temperature	Condition
Compressor in				
Compressor out				
Condenser in				
Condenser out				
Evaporator in				
Evaporator out				

MEASURING SYSTEM CAPACITY WITH A HEAT PUMP

LABORATORY OBJECTIVE

You will demonstrate your ability to determine the system capacity of an operating water source heat pump by measuring its water flow and temperature rise.

LABORATORY NOTES

In this lab we will calculate the capacity of an operating water source heat pump system by measuring the water flow rate and the water temperature difference. A Btu is defined as a 1°F temperature rise in 1 lb of water at atmospheric pressure. We can calculate the actual capacity accurately by multiplying the water flow in pounds per hour times the temperature rise in degrees Fahrenheit.

FUNDAMENTALS OF HVAC/R TEXT REFERENCE
Units 7 and 18

Required Tools and Equipment

Temperature sensor with temperature clamp and thermocouple clamp	Unit 12
Water source heat pump	Unit 49
1 gallon bucket	
Timer or watch that can measure seconds	

SAFETY REQUIREMENTS

Always familiarize yourself with the equipment and operating manuals prior to starting up any system.

PROCEDURE

STEP 1. Trace out the system and make sure that you understand the operation of the heat pump.

    A. Start the water loop through the heat pump before starting the unit.

    B. After water flow has been established, you may start the heat pump and run it in the cooling mode for 15 min.

    C. Measure the flow rate of the water. This can be done by allowing the water leaving the heat pump to flow into the 1 gal bucket. Time the duration in seconds from the time the bucket is initially empty until it is full.

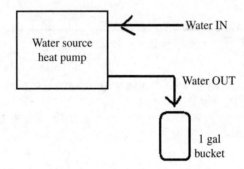

Figure P-9-1

Repeat this procedure for three different sets of readings.

    a) First measured time in seconds _____

    b) Second measured time in seconds _____

    c) Third measured time in seconds _____

    D. The average time in seconds is equal to the three readings added together and then divided by three.

    Average time in seconds = (Time 1 + Time 2 + Time 3) / 3

    Average time in seconds = _____

    E. To determine how many gallons per minute are flowing, first convert the seconds to minutes. Take the average time in seconds and divide that value by 60 (this is because there are 60 s in 1 min).

    Gallons per minute = total seconds / 60

    Gallons per minute (gpm) = _____

    F. Convert gallons per minute (gpm) to pounds per minute by multiplying gpm by 8.34 (this is because there are 8.34 lb of water in one gallon).

    Pounds per minute = gpm × 8.34

    Pounds per minute = _____

G. Measure the temperature of the water going in to the heat pump and the temperature of the water leaving the heat pump.

Temperature of the water IN _____

Temperature of the water OUT _____

H. Calculate the temperature difference, which is equal to the temperature OUT minus the temperature IN ($°F_{OUT} - °F_{IN}$).

Temperature difference (ΔT) = _____

I. Calculate the system capacity. To do this, multiply the pounds per minute from Step 1, Part F by the temperature difference from Step 1, Part H. This will then be multiplied by the specific heat of water at atmospheric pressure, which is simply 1 (1 Btu for every °F for every 1 lb of water). The calculated value will be in Btu per minute.

(lb/min $\times$ ($°F_{OUT} - °F_{IN}$) $\times$ 1 Btu/lb-°F) = Btu/min

System capacity in Btu/min = _____

J. Calculate the value for tons of refrigeration.

Remember that one ton of refrigeration is the equivalent of melting one ton (2,000 lb) of ice in 24 hr. Remember that the latent heat of fusion (ice to water) is 144 Btu/lb. An amount of 2,000 lb of ice multiplied by 144 Btu/lb is equal to 288,000 Btu for one ton in 24 hr. Divide this by 24 hr per day and you have 12,000 Btu/hr. Divide this by 60 min per hour and you have 200 Btu/min.

Therefore a ton of system capacity is equal to:
1 ton = 288,000 Btu/day
1 ton = 12,000 Btu/hr
1 ton = 200 Btu/min

To find capacity in tons, divide the system capacity in Btu/min by 200 Btu/min-ton.

System capacity in tons = (Btu/min)/200 Btu/min-ton

System capacity in tons = _____

K. How does the system capacity that you calculated compare to the name plate rating of the heat pump? Is the heat pump operating at full capacity?

_____

_____

_____

_____

STEP 2. Reverse the heat pump so that now it is in the heating mode and run it in this mode for 15 min before recording your next measurements.

A. Measure the flow rate of the water. Repeat this procedure for three different sets of readings.

First measured time in seconds _____

Second measured time in seconds _____

Third measured time in seconds _____

B. The average time in seconds is equal to the three readings added together and then divided by three.

Average time in seconds = (Time 1 + Time 2 + Time 3) / 3

Average time in seconds = _____

C. To determine how many gallons per minute are flowing, first convert the seconds to minutes. Take the average time in seconds and divide that value by 60 (this is because there are 60 s in 1 min).

Gallons per minute = total seconds / 60

Gallons per minute (gpm) = _____

D. Convert gallons per minute (gpm) to pounds per minute by multiplying gpm by 8.34 (this is because there are 8.34 lb of water in 1 gal).

Pounds per minute = gpm × 8.34

Pounds per minute = _____

E. Measure the temperature of the water going in to the heat pump and the temperature of the water leaving the heat pump.

Temperature of the water IN _____

Temperature of the water OUT _____

F. Calculate the temperature difference, which is equal to the temperature OUT minus the temperature IN ($°F_{OUT} - °F_{IN}$).

(Note: The temperature OUT minus the temperature IN will have a negative value. This is because in the heating mode the heat pump is absorbing heat from the water rather than rejecting heat. You do not need to use a negative value because you are just calculating the temperature difference; so you may ignore the negative sign for this experiment.)

Temperature difference (ΔT) = _____

G. Calculate the system capacity. To do this multiply pounds per minute from Step 2, Part D by the temperature difference from Step 2, Part F. This will then be multiplied by the specific heat of water at atmospheric pressure, which is simply 1 (one Btu for every °F for every 1 lb of water). The calculated value will be in Btu per minute.

[lb/min × ($°F_{OUT} - °F_{IN}$) × 1 Btu/lb-°F] = Btu/min

System capacity in Btu/min = _____

H. Calculate the value for tons of refrigeration.

Remember that one ton of refrigeration is the equivalent of melting one ton (2,000 lb) of ice in 24 hours. Remember that the latent heat of fusion (ice to water) is 144 Btu/lb. An amount of 2,000 lb of ice multiplied by 144 Btu/lb is equal to 288,000 Btu for one ton in 24 hours. Divide this by 24 hr per day and you have 12,000 Btu/hr. Divide this by 60 min per hour and you have 200 Btu/min.

Therefore a ton of system capacity is equal to:
1 ton = 288,000 Btu/day
1 ton = 12,000 Btu/hr
1 ton = 200 Btu/min

To find capacity in tons, divide the system capacity in Btu/min by 200 Btu/min-ton.

System capacity in tons = (Btu/min) / 200 Btu/min-ton

System capacity in tons = _____

I. How does the system capacity that you calculated compare to the name plate rating of the heat pump? Is the heat pump operating at full capacity?

_____

_____

_____

_____

J. How does the system capacity when operating the heat pump in the cooling mode differ from running the heat pump in the heating mode? Explain your answer.

_____

_____

_____

_____

ENERGY TYPES

LABORATORY OBJECTIVE

You will demonstrate that energy can be converted from one form to another and that electricity generated from a rotating magnetic device is not free but rather this takes work input.

LABORATORY NOTES

We will use a hand cranked electric generator that is connected to several electric lightbulbs. The lights will be controlled by switches. You will turn the generator by hand, producing electricity. You will control the lights with switches to demonstrate the effect of electrical loads on the generator.

FUNDAMENTALS OF HVAC/R TEXT REFERENCE
Unit 28

Required Tools and Equipment

Hand cranked generator	
3 Electric single pole, single throw switches	
3 Lights	

SAFETY REQUIREMENTS

    A. Check all circuits for voltage before doing any service work.

    B. Stand on dry nonconductive surfaces when working on live circuits.

    C. Never bypass any electrical protective devices.

PROCEDURE

STEP 1. Familiarize yourself with the generator and the electrical circuit.

    A. Make sure that all of the spst (single pole, single throw) switches are open (off).

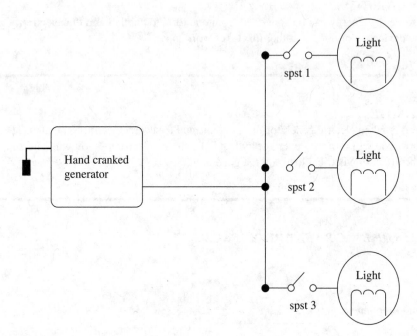

Figure P-10-1

    B. Begin turning the generator hand crank. This is the amount of energy required to overcome the mechanical inefficiencies in the machine.

    C. While turning the generator, switch on (close) the first spst switch to turn on the smallest light. Notice that the generator becomes harder to turn.

    D. Continue turning the generator hand crank while switching on the other lights one at a time. Notice the change in the energy required to turn the hand crank.

    E. Continue turning the generator hand crank while switching off the other lights one at a time. Notice the change in the energy required to turn the hand crank.

QUESTIONS

1. What types of energy are represented in this experiment?

2. What change occurs in the generator each time an electric load is added?

3. What change occurs in the generator each time an electrical load is removed?

4. Why does this happen?

BASIC REFRIGERATION SYSTEM STARTUP

LABORATORY OBJECTIVE
The student will demonstrate the correct procedure for installing a gauge manifold to read system operating pressures when conducting routine preventive maintenance.

LABORATORY NOTES
This basic startup assumes the system had been working in the past and is still expected to be operable but has not been used for several months. This procedure is similar to the maintenance you would perform on an ice cream parlor that has been closed for the winter and will open for the summer, or a school that has been closed for the summer. It is your job to perform the prestart visual inspection, install the gauges, check system idle pressures, and to operate the system in a normal mode, making a judgment as to current system operating condition.

FUNDAMENTALS OF HVAC/R TEXT REFERENCE
Unit 36

Required Tools and Equipment

Gloves and goggles	Unit 3
Service valve wrench	Unit 9
Gauge manifold	Unit 12
Operating refrigeration system	

SAFETY REQUIREMENTS

A. Wear safety goggles and gloves when working with refrigerants. Liquid refrigerant can cause frostbite when in contact with eyes and skin.

B. Use low loss hose fittings, or wrap cloth around hose fittings before removing the fittings from a pressurized system or cylinder. Inspect all fittings before attaching hoses.

PROCEDURE

STEP 1. Familiarize yourself with the major components in the refrigeration system including the condenser, compressor, evaporator, and metering device. Determine where the high side and low side connections for the system are located.

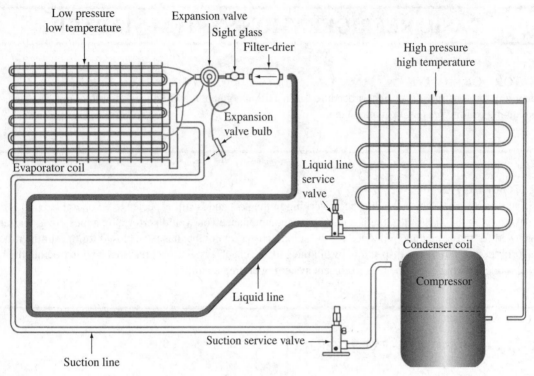

Figure R-1-1

STEP 2. The refrigerant type and required amount of charge can be found on the refrigeration system name plate.

F.L.A.   F.L.A.

L.R.A.   L.R.A.

H.P.   H.P.

Volts   Volts

Hertz   Hertz
phase   phase

R-22   5.0   lb   kg

Figure R-1-2

STEP 3. Determine the minimum operating head pressure. Use a pressure-temperature (P-T) relationship chart such as shown in Table R-1-1 and from Figure 23-2 in Unit 23 of the *Fundamentals of HVAC/R* text.

TABLE R-1-1 Vapor Pressure/Temperature of Refrigerants

Temp (Deg F)	CFC R-11	CFC R-12	CFC R-113	CFC R-114	CFC R-500	CFC R-502	HCFC R-22	HCFC R-123	HCFC R-124	HFC R-125	HFC R-134a	HFC R-410a
-100	29.8	27.0			26.4	23.3	25.0	29.9	29.2	24.4	27.8	
-90	29.7	25.7			24.9	20.6	23.0	29.8	28.8	21.7	26.9	
-80	29.6	24.1			22.9	17.2	20.2	29.7	28.2	18.1	25.6	
-70	29.4	21.8			20.3	12.8	16.6	29.6	27.4	13.3	23.8	
-60	29.2	19.0			17.0	7.2	12.0	29.5	26.3	7.1	21.5	
-50	28.9	15.4			12.8	0.2	6.2	29.2	24.8	0.3	18.5	5.8
-40	28.4	11.0			7.6	4.1	0.5	28.9	22.8	4.9	14.7	11.7
-30	27.8	5.4			1.2	9.2	4.9	28.5	20.2	10.6	9.8	18.9
-20	27.0	0.6	29.0	22.8	3.2	15.3	10.2	27.8	16.9	17.4	3.8	27.5
-10	26.0	4.4	28.6	20.5	7.8	22.6	16.4	27.0	12.7	25.6	1.8	38.8
0	24.7	9.2	28.1	17.7	13.3	31.1	24.0	26.0	7.6	35.1	6.3	49.8
10	23.1	14.6	27.5	14.3	19.7	41.0	32.8	24.7	1.4	46.3	11.6	63.9
20	21.1	21.0	26.7	10.1	27.2	52.4	43.0	23.0	3.0	59.2	18.0	80.2
30	18.6	28.4	25.7	5.1	36.0	65.6	54.9	20.8	7.5	74.1	25.6	99.0
40	15.6	37.0	24.4	0.4	46.0	80.5	68.5	18.2	12.7	91.2	34.5	120.0
50	12.0	46.7	22.9	3.9	57.5	97.4	84.0	15.0	18.8	110.6	44.9	144.9
60	7.8	57.7	20.9	7.9	70.6	116.4	101.6	11.2	25.9	132.8	56.9	172.5
70	2.8	70.2	18.6	12.6	85.3	137.6	121.4	6.6	34.1	157.8	70.7	203.6
80	1.5	84.2	15.8	18.0	101.9	161.2	143.6	1.1	43.5	186.0	86.4	238.4
90	4.9	99.8	12.4	24.2	120.4	187.4	168.4	2.6	54.1	217.5	104.2	277.3
100	8.8	117.2	8.5	31.2	141.1	216.2	195.9	6.3	66.2	252.7	124.3	320.4
110	13.1	136.4	3.8	39.1	164.0	247.9	226.4	10.5	79.7	291.6	146.8	368.2
120	18.3	157.7	0.8	48.0	189.2	282.7	259.9	15.4	94.9	334.3	171.9	420.9
130	24.0	181.0	3.8	58.0	217.0	320.8	296.8	21.0	111.7	380.3	199.8	478.9
140	30.4	206.6	7.3	69.1	247.4	362.6	337.2	27.3	130.4	430.2	230.5	542.5
150	37.7	234.6	11.2	81.4	280.7	408.4	381.5	34.5	151.0	482.1	264.4	612.1
160								42.5	173.6		301.5	684.0
170								51.5	198.4		342.0	
180								61.4	225.6		385.9	
190								72.5	255.1		433.6	
200								84.7	287.3		485.0	
210								98.1	322.1		540.3	
220								112.8	359.9			
230								128.9	400.6			
240								146.3	444.5			
250								165.3	491.8			

italics are in inches of mercury vacuum (in/Hg or " Hg)

bold are in pounds per square inch pressure (psig)

A. A rule of thumb for checking the high side for proper operation is to take a temperature reading of the ambient air temperature. In order to maintain a proper condensing temperature at the condenser, there must be approximately a 20–35°F difference in ambient temperature and condensing temperature.

B. **EXAMPLE:** For an 80°F ambient temperature and a 20°F temperature difference using refrigerant R-410a, the condensing temperature is 100°F. Looking at the P-T chart, Table R-1-1, this means that the high side gauge should read at least 320.4 psig pressure.

C. In analyzing the low side (compound gauge) pressure, suppose the unit is designed for an evaporator temperature of 30°F at 65°F ambient and 40°F at 85°F ambient. Using R-410A, the pressure of boiling refrigerant in the evaporator at 30°F would be 99.0 psig and at 40°F it would be 120.0 psig.

STEP 4. Connect the gauge manifold using the procedure provided below.

A. Remove the valve stem caps as shown in Figure R-1-3 from the equipment service valves and check to be sure that both service valves are back seated (turned all the way out, counterclockwise).

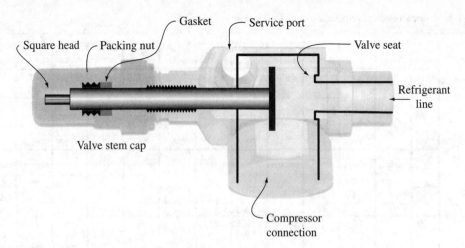

Figure R-1-3

B. Remove the service (gauge) port caps from both service valves.

C. Connect the center hose from the gauge manifold to a refrigerant cylinder, using the same type of refrigerant that is in the system and open both valves on the gauge manifold.

D. Open the valve on the refrigerant cylinder for about 2 s and then close it. This will purge any contaminants from the gauge manifold and hoses as shown in Figure R-1-4.

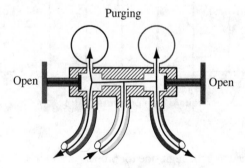

Figure R-1-4

E. Connect the gauge manifold hoses to the gauge ports—the low pressure compound gauge to the suction service valve and the high pressure gauge to the high side service valve as shown in Figure R-1-5.

F. Purge the air from each hose one at a time. This is done by first opening the valve on the refrigerant cylinder. Then slightly loosen the hose connection at the suction line service valve for about 2 s. You will hear the line purging and then tighten the connection. Repeat this with for the high side service valve connection.

G. After the lines are purged of air, then close the cylinder valve and front seat (close) both valves on the gauge manifold as shown in Figure R-1-6. Crack (turn clockwise) both service valves one turn off the back seat. The system pressure is now allowed to register on each gauge.

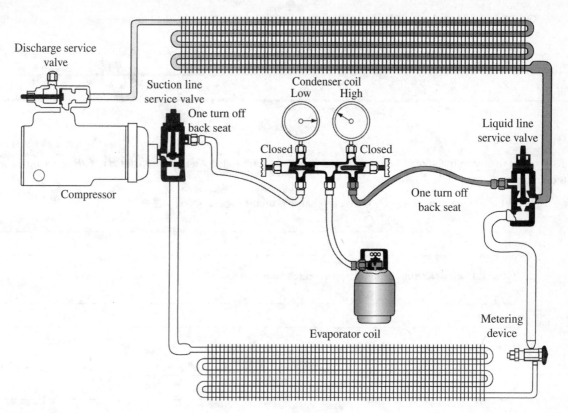

Figure R-1-5

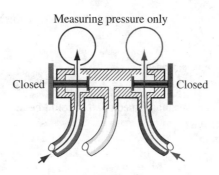

Measuring pressure only

Closed — Closed

Figure R-1-6

STEP 5. Record the idle pressures and then turn the system on. Observe the high side pressure go up and the low side pressure go down. Allow 5 min of operation for the system to obtain normal pressures and then record the following.

Idle
High side pressure _____   Low side pressure _____

Running
High side pressure _____   Low side pressure _____

STEP 6. Once the readings are recorded, prepare to remove the gauge manifold from the system using the following procedure.

A. Back seat the liquid line first (counterclockwise).

B. Mid seat the valves on the gauge manifold to allow the remaining liquid in the high pressure hose to be pulled into the suction side of the system.

113

C. After the pressures equalize, back seat the suction service valve on the compressor.

D. Remove the hoses from the gauge ports and seal the ends of the hoses by reattaching them to the back of the gauge manifold set.

QUESTIONS

To get help in answering some of the following questions, refer to the *Fundamentals of HVAC/R* text Unit 36 and also the Troubleshooting Table 36-4.

1. What are the expected suction and discharge pressures for this system?

   _____

   _____

2. How did you determine what these pressures should be?

   _____

   _____

3. How do the pressures you recorded compare with the expected pressures?

   _____

   _____

(Circle the letter that indicates the correct answer.)

4. If the discharge pressure is higher than expected, this could be due to:
   A. a dirty condenser.
   B. bad compressor bearings.
   C. excessive compressor lubrication.
   D. an inadequate supply of refrigerant.

5. If the discharge pressure is lower than expected, this could be due to:
   A. moisture in the refrigerant.
   B. air trapped in the system.
   C. a low charge of refrigerant.
   D. insufficient condenser cooling.

6. If the suction pressure is lower than expected, this could be due to:
   A. excessive condenser cooling.
   B. the expansion valve not opening.
   C. an excessive charge of refrigerant.
   D. air trapped in the evaporator.

7. When service valves are back seated:
   A. the gauge port is closed to system pressure.
   B. the gauge port is open to system pressure.
   C. there will be no refrigerant flow through the valve.
   D. The entire valve is closed.

8. Whenever purging air from gauge manifold hoses:
   A. be sure to minimize the amount of refrigerant release.
   B. never use the same refrigerant as that found in the system.
   C. always use a refrigerant with a higher boiling point to purge.
   D. always use a refrigerant with a lower boiling point to purge.

9. Low loss gauge manifold hose fittings:
   A. reduce the amount of refrigerant lost.
   B. prevent the refrigerant from being released and causing skin burns.
   C. Both A and B are correct.
   D. Neither A nor B is correct.

10. An overcharge of refrigerant would be indicated by:
    A. a lower than expected discharge pressure.
    B. a higher than expected discharge pressure.
    C. a lower than expected suction pressure.
    D. a suction pressure of exactly 0 psig.

11. A worn compressor could be indicated by:
    A. a very cold evaporator.
    B. a very hot condenser.
    C. a chattering expansion valve.
    D. a high suction pressure and low discharge pressure.

12. Precautions to take when using a gauge manifold include:
    A. never dropping or abusing the gauge manifold.
    B. keeping the ports or charging lines capped when not in use.
    C. never using any fluid other than clean oil and refrigerant.
    D. All of the above are correct.

13. If the high and low pressure readings on a gauge manifold are identical:
    A. both valves on the gauge manifold are open.
    B. both valves on the gauge manifold are shut.
    C. the compressor is short cycling.
    D. the condenser is dirty.

14. Under EPA regulations, air conditioning technicians are allowed to purge the gauge hoses provided it is a de minimis release.
    A. True.
    B. False.

15. A gauge manifold can measure:
    A. pressure and volume only.
    B. pressure only, along with corresponding saturation temperature.
    C. both pressure and temperature.
    D. temperature only.

REFRIGERANT VAPOR RECOVERY SELF-CONTAINED ACTIVE

LABORATORY OBJECTIVE

The student will demonstrate the correct procedure for the self-contained active method of recovering vapor refrigerant from a refrigeration system.

LABORATORY NOTES

This lab is intended to help students practice the removal of refrigerant from a system using the self-contained active method of vapor recovery. Refrigerant recovery must be done on any and all systems prior to any refrigerant side repairs or dismantling.

FUNDAMENTALS OF HVAC/R TEXT REFERENCE

Unit 26

Required Tools and Equipment

Gloves and goggles	Unit 3
Recovery cylinder and hoses	Unit 26
Recovery unit	Unit 26
Refrigerant scale	Unit 27
Disabled refrigeration system	

SAFETY REQUIREMENTS

A. Wear safety goggles and gloves when working with refrigerants. Liquid refrigerant can cause frostbite when in contact with eyes and skin.

B. Use low loss hose fittings, or wrap cloth around hose fittings before removing the fittings from a pressurized system or cylinder. Inspect all fittings before attaching hoses.

PROCEDURE

STEP 1. Familiarize yourself with the refrigerant recovery unit that will be used for the lab and carefully read the instructions that came with the unit. It is important to remember that the recovery unit utilizes its own built in compressor (self-contained). Compressors can only pump vapors. Therefore liquid refrigerant must NEVER be introduced into the recovery unit!

STEP 2. Connect and tighten both ends of a refrigerant hose from the disabled refrigeration system to the recovery unit. Include a filter-drier in the line if available.

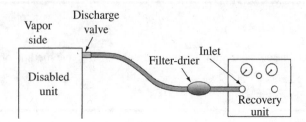

Figure R-2-1

STEP 3. Familiarize yourself with the refrigerant recovery cylinder.

    A. Recovery cylinders must be hydrostatically tested every five years and are stamped on the neck with an expiration date. Recovery cylinders have a yellow neck and gray body and new cylinders are either sealed in a vacuum or with an inert gas. NEVER mix refrigerants! Once the cylinder is first used, the refrigerant type should be clearly marked on the yellow neck with a permanent marker.

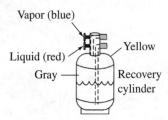

    Figure R-2-2

    B. The recovery cylinder has two valves, the blue vapor valve and the red liquid valve. Red indicates danger as liquid refrigerant has a greater potential to cause frostbite to eyes and exposed skin.

    C. Recovery cylinders must never be filled to a level of more than 80% full. Refrigerants are always measured by weight. Stamped on the neck of the cylinder you will find the tare weight (T.W.) and the water capacity (W.C.). The maximum amount of refrigerant that the cylinder can hold is equal to $0.8 \times$ W.C.

    D. The tare weight is the weight of an empty cylinder. If you are weighing a recovery cylinder on a refrigerant scale then you need to include this in your calculation. Therefore the total maximum weight of any recovery cylinder is equal to $0.8 \times$ W.C. + T.W.

STEP 4. Connect and purge the lines as follows:

    A. Tighten both ends of a refrigerant hose from the outlet of the recovery unit to the vapor valve (blue) on the recovery cylinder and place the cylinder on top of a refrigerant scale.

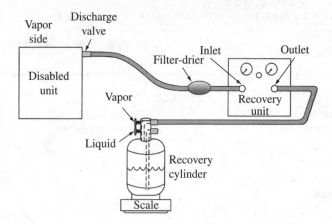

Figure R-2-3

B. Crack open the discharge of the disabled unit and purge air from the refrigerant hose by loosening the fitting at the recovery unit for 2 s. Then open the disabled unit discharge valve fully.

C. Crack open the vapor valve (blue) on the recovery cylinder and purge air from the refrigerant hose by loosening the fitting at the recovery unit for 2 s. Then open the vapor valve on the recovery cylinder fully.

STEP 5. Proceed to start the recovery unit according to the manufacturer's instructions.

A. The recovery unit may have a bypass switch that allows it to start without a load.

B. It may also have a built in condenser fan, which may need to be switched on.

C. You will be drawing vapor into the recovery unit where it will be passed through a condenser. Therefore liquid refrigerant will discharge from the recovery unit and pass through the connected hose to the vapor valve on the recovery cylinder and drop from the vapor connection into the remaining liquid at the bottom.

D. You are recovering vapor; however, you will be sending liquid to the recovery cylinder.

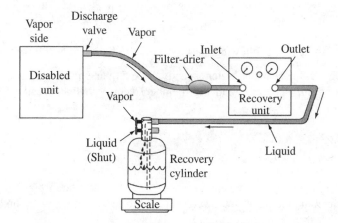

Figure R-2-4

STEP 6. The recovery unit is equipped with a low pressure cutout.

A. These controls are usually set below 29 in Hg for low pressure refrigerants and at 10 in Hg for high pressure refrigerants.

TABLE R-2-1 Required Levels of Evacuation for High Pressure Appliances

Recover Equipment Refrigerant and Charge	Manufacturing Date	
	Before 11/15/98 (in Hg)	After 11/15/98 (in Hg)
R-22 Appliance, <200 lb charge	0*	0
R-22 Appliance, >200 lb charge	4	10
Other high pressure appliance <200 lb charge	4	10
Other high pressure appliance >200 lb charge	4	15
Very high pressure equipment	0	0

*A zero (0) vacuum is atmospheric pressure. A perfect vacuum is 30 in Hg.

B. When the machine shuts off on the low pressure cutout, the recovery is not necessarily completed.

C. If the machine remains idle for a short period of time, the pressure may creep up. The recovery machine should be run again, this time with the condenser fan off if possible. This will allow the vapor to push any remaining liquid out of the hose connected to the recovery cylinder, thereby minimizing any refrigerant loss when you disconnect.

D. Usually when the machine cuts off twice, the recovery is considered complete.

E. Laboratory Manual Table R-2-1 and Figure 26-3 of the *Fundamentals of HVAC/R* text list the EPA requirements.

STEP 7. After the recovery is complete, close the valve on the disabled unit and the vapor valve on the recovery cylinder and carefully disconnect the hoses.

QUESTIONS

To get help in answering some of the following questions, refer to the *Fundamentals of HVAC/R* text Unit 26.

1. Given a recovery cylinder with a water capacity of 26.5 lb and a tare weight of 14.6 lb, calculate the maximum weight of a full recovery cylinder (remember: recovery cylinders can only be filled to 80% capacity).

2. Given a recovery cylinder with a water capacity of 32.6 lb and a tare weight of 15.7 lb, the cylinder already has some refrigerant in it and its total weight is 27.3 lb. How many more pounds of refrigerant will the cylinder be able to hold?

3. The active method of refrigerant recovery:
   A. uses the small appliance compressor to recover the refrigerant.
   B. uses the pressure in the system to recover the refrigerant.
   C. uses self-contained recovery equipment.
   D. can be used on high pressure systems only.

4. The recovery cylinder:
   A. must have DOT approval.
   B. can never be refilled.
   C. can only be filled to 95% capacity.
   D. is color coded blue.

5. Very high pressure equipment contains refrigerant that:
   A. has a pressure above 1000 psig.
   B. has a pressure in excess of 152 psig.
   C. has a boiling point below –50°C (–58°F) at atmospheric pressure.
   D. can never be recovered due to excessive pressure.

6. A gray cylinder with a yellow top:
   A. is color coded for a disposable cylinder.
   B. is always used with propane.
   C. is always used with nitrogen.
   D. is color coded for a recovery cylinder.

7. Whenever a refrigeration circuit needs to be opened up for service:
   A. if there is still any refrigerant in the system, the refrigerant charge must be recovered before the system is opened.
   B. you can simply release all remaining refrigerant to the atmosphere.
   C. notify the EPA before servicing.
   D. you can only release the refrigerant vapor to the atmosphere.

8. The date stamped on the neck of a recovery cylinder:
   A. indicates when it was manufactured.
   B. indicates the last date that the cylinder can be used for recovering, storing, or handling refrigerants.
   C. Both A and B are correct.
   D. Neither A nor B is correct.

9. In an emergency, a disposable cylinder can be used as a replacement for a recovery cylinder.
   A. True.
   B. False.

10. When brand new, recovery cylinders are 80% full of refrigerant.
    A. True.
    B. False.

11. Recovery cylinders:
    A. can be used for mixtures of refrigerants.
    B. should be used to hold one refrigerant type only.
    C. must always be stored laying on their sides.
    D. Both A and C are correct.

12. When a refrigerant is first introduced into a recovery cylinder:
    A. the cylinder should be clearly marked with the refrigerant designation using a permanent marker.
    B. a tag should be placed around the cylinder valve.
    C. a piece of tape should be placed on the bottom of the cylinder.
    D. nothing needs to be done.

13. When brand new, recovery cylinders:
    A. may have a press charge of an inert gas.
    B. may be under a vacuum.
    C. Both A and B are correct.
    D. None of the above is correct.

14. Recovery cylinders have two valves:
    A. and one is the inlet while the other is the outlet.
    B. and one is for refrigerant while the other is for an inert gas.
    C. and one is for liquid while the other is for vapor.
    D. None of the above is correct.

15. A recovery cylinder liquid valve:
    A. is usually red.
    B. is connected to an inner tube that extends to the bottom of the cylinder.
    C. allows for access to the liquid without needing to turn the cylinder upside down.
    D. All of the above are correct.

LIQUID RECOVERY SELF-CONTAINED ACTIVE

LABORATORY OBJECTIVE

The student will demonstrate the correct procedure for the self-contained active method of recovering liquid refrigerant from a refrigeration system. The student should review Lab R-2 before beginning this exercise.

LABORATORY NOTES

This lab is intended to help students practice the removal of refrigerant from a system using the self-contained active method of liquid recovery. The liquid is forced out of the disabled unit using the recovery machine to lower the pressure in the recovery cylinder and increase the pressure in the disabled unit. This causes rapid movement of the liquid. The advantage of the liquid recovery method of transfer is that liquid recovery is much faster than the vapor recovery method. However the final evacuation still must be done by the vapor recovery method (Lab R-2).

FUNDAMENTALS OF HVAC/R TEXT REFERENCE
Unit 26

Required Tools and Equipment

Gloves and goggles	Unit 3
Recovery cylinder and hoses	Unit 26
Recovery unit	Unit 26
Refrigerant scale	Unit 27
Disabled refrigeration system	

SAFETY REQUIREMENTS

A. Wear safety goggles and gloves when working with refrigerants. Liquid refrigerant can cause frostbite when in contact with eyes and skin.

B. Use low loss hose fittings, or wrap cloth around hose fittings before removing the fittings from a pressurized system or cylinder. Inspect all fittings before attaching hoses.

PROCEDURE

STEP 1. Familiarize yourself with the refrigerant recovery unit that will be used for the lab and carefully read the instructions that came with the unit. It is important to remember that the recovery unit utilizes its own built in compressor (self-contained). Compressors can only pump vapors. Therefore liquid refrigerant must NEVER be introduced into the recovery unit!

STEP 2. Connect and tighten both ends of a refrigerant hose from the vapor side of the disabled refrigeration system to the recovery unit.

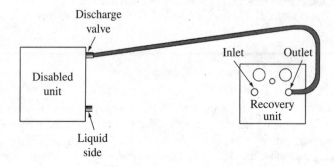

Figure R-3-1

STEP 3. Familiarize yourself with the refrigerant recovery cylinder. For additional information on the use of recovery cylinders refer to Lab R-2 *Vapor Recovery Self-Contained Active,* Step 3.

STEP 4. Connect and tighten both ends of a refrigerant hose from the inlet of the recovery unit to the vapor valve (blue) on the recovery cylinder and place the cylinder on top of a refrigerant scale.

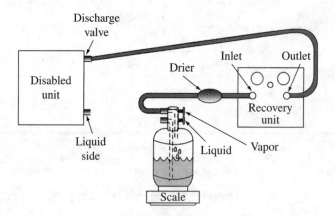

Figure R-3-2

STEP 5. Connect and tighten both ends of a refrigerant hose including a sight glass from the liquid side of the disabled unit to the liquid valve (red) on the recovery cylinder.

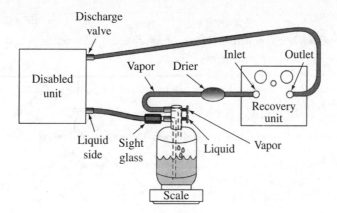

Figure R-3-3

STEP 6. Purge the lines of air as follows:

  A. Crack open the discharge of the disabled unit and purge air from the refrigerant hose by loosening the fitting at the recovery unit for 2 s. Then open the disabled unit discharge valve fully.

  B. Crack open the vapor valve (blue) on the recovery cylinder and purge air from the refrigerant hose by loosening the fitting at the recovery unit for 2 s. Then open the vapor valve on the recovery cylinder fully.

  C. Crack open the liquid valve (red) on the recovery cylinder and purge air from the refrigerant hose by loosening the fitting at the disabled unit for 2 s. Then open the disabled unit liquid side valve fully.

STEP 7. Proceed to start the recovery unit according to the manufacturer's instructions.

  A. The recovery unit may have a bypass switch that allows it to start without a load.

  B. It may also have a built in condenser fan, which should remain OFF.

  C. You will be drawing vapor into the recovery unit where its pressure will be increased and sent to the disabled unit. The vapor entering the disabled unit will push the liquid through the sight glass and into the recovery cylinder (push).

  D. Continue this process until the entire quantity of liquid refrigerant has been removed from the disabled unit and no more liquid appears through the sight glass.

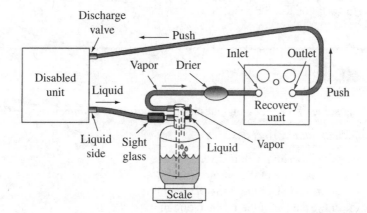

Figure R-3-4

STEP 8.

A. After all of the liquid has been removed, close the vapor valve (blue) on the recovery cylinder and the recovery unit will stop on the low pressure cutout.

B. When the recovery unit has stopped, close the discharge valve and the liquid side valve on the disabled unit and the liquid valve (red) on the recovery cylinder.

C. Carefully disconnect the hoses and then reinstall them to proceed with a vapor recovery (pull) as described in Laboratory Worksheet R-2.

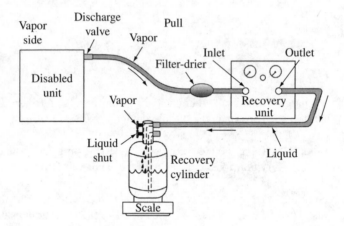

Figure R-3-5

STEP 9. When the machine shuts off on the low pressure cutout, the recovery is not necessarily completed.

A. If the machine remains idle for a short period of time, the pressure may creep up. The recovery machine should be run again.

B. This time run the recovery machine with the condenser fan off if possible. This will allow the vapor to push any remaining liquid out of the hose connected to the recovery cylinder, thereby minimizing any refrigerant loss when you disconnect.

C. After the recovery is complete, close the valve on the disabled unit and the vapor valve on the recovery cylinder and carefully disconnect the hoses.

QUESTIONS

To get help in answering some of the following questions, refer to the *Fundamentals of HVAC/R* text Unit 26.

(Circle the letter that indicates the correct answer.)

1. The liquid recovery procedure will recover about:
   A. 45% of the entire charge of the refrigeration unit.
   B. 95% of the entire charge of the refrigeration unit.
   C. 75% of the entire charge of the refrigeration unit.
   D. the entire charge of the refrigeration unit.

2. A liquid recovery:
   A. is always followed by a vapor recovery.
   B. is followed by a vapor recovery about 50% of the time.
   C. directs liquid into the inlet of the recovery unit.
   D. is slower than a vapor recovery.

3. To recover the refrigerant from a centrifugal chiller:
   A. always pull all the vapor first.
   B. always push out all of the liquid first.
   C. always push out all of the vapor first.
   D. None of the above is correct.

4. Pulling vapor from a centrifugal chiller before removing the liquid:
   A. is a proper method.
   B. is extremely fast.
   C. is always recommended.
   D. can damage the chiller tubes due to the chill water freezing up.

5. When recovering refrigerant from a centrifugal chiller:
   A. the chill water pump should be running to reduce the possibility of freeze-up.
   B. the chill water may be warmed slightly to reduce the possibility of freeze-up.
   C. the push–pull method should be used.
   D. All of the above are correct.

6. When performing a liquid recovery:
   A. a sight glass should be installed in the discharge line of the recovery unit.
   B. a drier should be installed in the discharge line of the recovery unit.
   C. a sight glass should be installed in the liquid line connected to the recovery cylinder.
   D. All of the above are correct.

LIQUID RECOVERY SYSTEM DEPENDENT PASSIVE

LABORATORY OBJECTIVE

The student will demonstrate the correct procedure for the system dependent passive method of recovering liquid refrigerant from a refrigeration system. The student should review the section on recovery cylinders from Laboratory Worksheet R-2 (Step 3) and connecting gauge manifolds from Laboratory Worksheet R-1 (Step 4) before beginning this exercise.

LABORATORY NOTES

This lab is intended to help students practice the removal of refrigerant from a system using the system dependent passive method of liquid recovery. The liquid is pumped out of the refrigeration unit by using its own compressor. The only recovery equipment required is a recovery cylinder. With this method, 90% of the refrigerant must be recovered.

FUNDAMENTALS OF HVAC/R TEXT REFERENCE
Unit 26

Required Tools and Equipment

Gloves and goggles	Unit 3
Recovery cylinder	Unit 26
Refrigerant scale	Unit 27
Service valve wrench	Unit 9
Gauge manifold	Unit 12
Operating refrigeration system	

SAFETY REQUIREMENTS

A. Wear safety goggles and gloves when working with refrigerants. Liquid refrigerant can cause frostbite when in contact with eyes and skin.

B. Use low loss hose fittings, or wrap cloth around hose fittings before removing the fittings from a pressurized system or cylinder. Inspect all fittings before attaching hoses.

PROCEDURE

STEP 1. Familiarize yourself with the major components in the refrigeration system including the condenser, compressor, evaporator, and metering device. Determine where the high side and low side connections for the system are located.

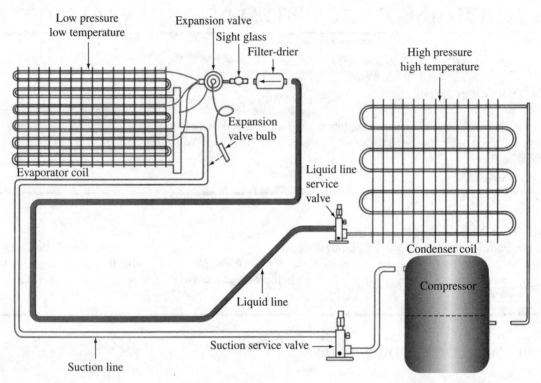

Figure R-4-1

STEP 2. Connect the gauge manifold as shown below and purge the lines of air using the procedure provided in Laboratory Worksheet R-1 (Step 4). The recovery cylinder hose should be connected to the liquid side (red) valve on the cylinder.

STEP 3. After the lines have been connected and purged, place the recovery cylinder on a scale. Familiarize yourself with the refrigerant recovery cylinder. For additional information on the use of recovery cylinders refer to Laboratory Worksheet R-2 *Vapor Recovery Self-Contained Active,* Step 3.

STEP 4. To prepare for recovery, line up the system as follows:

   A. The liquid line service valve should be in the mid position to allow flow through the service port.

   B. The high side gauge manifold valve should be closed.

   C. The liquid valve (red) on the recovery cylinder should be wide open.

   D. The low side gauge manifold valve should be closed.

   E. The suction line service valve should be open one turn off its back seat.

130

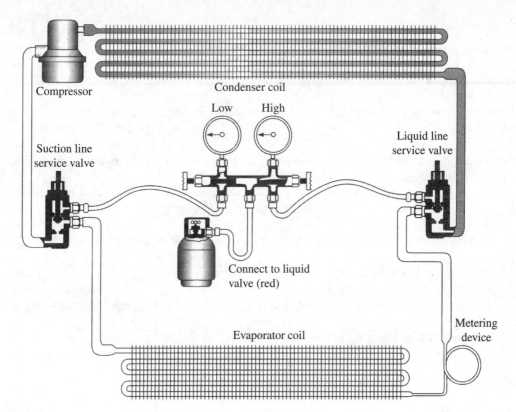

Figure R-4-2

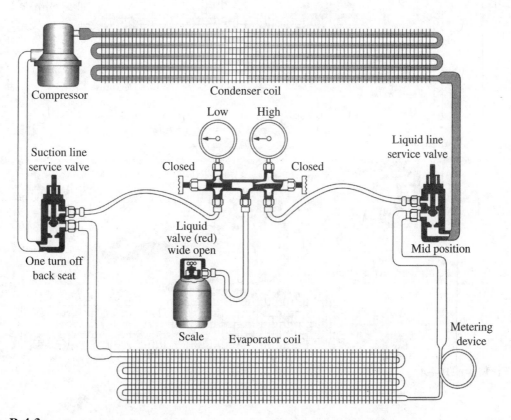

Figure R-4-3

STEP 5. Begin recovery as follows:

A. Start condenser cooling fan or cooling water flow as normal.

B. Start compressor and monitor the high side and low side pressures.

C. Slowly open the gauge manifold high side valve wide open and refrigerant should begin flowing into the recovery cylinder.

D. Do not allow discharge pressure to exceed the normal recovery cylinder maximum pressure (usually no greater than 250 psig).

E. Monitor the cylinder weight (remember no more than 80% full).

F. Cool the recovery cylinder with cold water or ice to help speed recovery if necessary.

G. When the low side pressure reaches 0 psig, the recovery is complete.

H. Close the liquid fill valve on the recovery cylinder and then shut down the refrigeration system.

I. Back seat the high side and low side service valves.

J. Carefully disconnect the hoses.

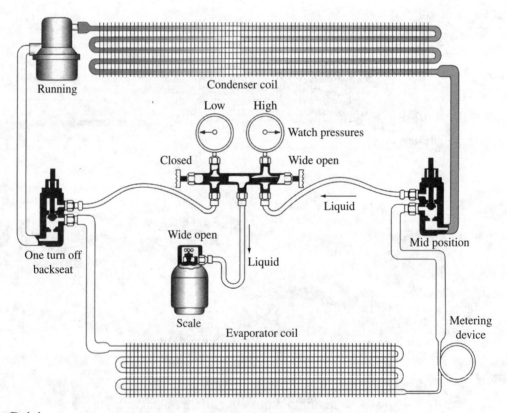

Figure R-4-4

QUESTIONS

To get help in answering some of the following questions, refer to the *Fundamentals of HVAC/R* text Unit 26.

(Circle the letter that indicates the correct answer.)

1. The system dependent passive method of recovery with the compressor running should be capable of recovering:
   A. 80% of the entire charge of the refrigeration unit.
   B. 90% of the entire charge of the refrigeration unit.
   C. 75% of the entire charge of the refrigeration unit.
   D. the entire charge of the refrigeration unit.

2. The system dependent passive method of recovery *without* the compressor running should be capable of recovering:
   A. 80% of the entire charge of the refrigeration unit.
   B. 90% of the entire charge of the refrigeration unit.
   C. 75% of the entire charge of the refrigeration unit.
   D. the entire charge of the refrigeration unit.

3. The push-pull method of recovery is considered to be:
   A. system dependent active.
   B. system dependent passive.
   C. self-contained passive.
   D. self-contained active.

4. If you do not have a recovery unit and the system compressor does not work:
   A. you do not have to recover the refrigerant.
   B. the "grandfather" clause allows you to simply vent the charge.
   C. try to run the compressor by hand.
   D. you still must recover the refrigerant.

5. If you recover liquid refrigerant using the system dependent passive method:
   A. you will still need to use an appropriate recovery cylinder.
   B. you can use any type of refrigerant cylinder.
   C. you must use a disposable refrigerant cylinder.
   D. None of the above is correct.

6. The method of recovery used in a system dependent passive procedure:
   A. is most closely related to push.
   B. is most closely related to push-pull.
   C. is most closely related to pull.
   D. All of the above are somewhat correct.

7. If you have no recovery unit and the system compressor will not operate:
   A. just cut the refrigerant lines to release the charge.
   B. cut the lines but be very careful not to get frostbite.
   C. always recover the required amount of refrigerant prior to opening the system.
   D. None of the above is correct.

8. If the system compressor does not operate then the only alternative is to remove 80% of the charge using the system dependent passive method of recovery.
   A. True.
   B. False.

DEEP METHOD OF EVACUATION

LABORATORY OBJECTIVE

The student will demonstrate the correct procedure for the deep method of evacuation of a refrigeration system.

LABORATORY NOTES

This laboratory worksheet is intended to help students practice the evacuation of a refrigeration system. Proper evacuation of a system will remove noncondensable gases (air, water vapor, and inert gases) and ensure a tight, dry system before charging.

FUNDAMENTALS OF HVAC/R TEXT REFERENCE
Unit 27

Required Tools and Equipment

Gloves and goggles	Unit 3
Deep vacuum pump and micron gauge	Unit 27
Service valve wrench	Unit 9
Gauge manifold	Unit 12
Refrigeration system	

SAFETY REQUIREMENTS

Wear safety goggles and gloves when working on refrigeration systems. Oil in the system can become acidic over time and can cause acid burns to the skin and eyes.

PROCEDURE

STEP 1. Familiarize yourself with the deep vacuum pump. It is somewhat like an air compressor in reverse.

    A. They are rated according to free air displacement in cubic feet per minute (cfm), or liters per minute (L/m) in the SI system.

B. The degree of vacuum the pump can achieve is expressed in microns. One micron is equivalent to 1/25,400 of an inch of mercury.

C. Deep vacuum pumps achieve levels of 500 microns or less and a reading of 0 microns would be equal to a perfect vacuum. These pressures are too small to read on a standard gauge manifold, therefore most deep vacuum pumps include a micron gauge as shown in Figure R-5-1.

Micron gauge

Figure R-5-1

D. To illustrate the measurement of microns, refer to Table R-5-1 shown below and Figure 27-8 of the *Fundamentals of HVAC/R* text.

E. Note: Table R-5-1 not only demonstrates the comparison in units of measure but dramatically shows the changes in the boiling point of water as the evacuation approaches the perfect vacuum.

F. This is the main purpose of evacuation—to reduce the pressure enough to boil or vaporize the water and then pump it out of the system.

TABLE R-5-1 Comparison of three different pressure measuring systems

Boiling point of water		Unit of absolute pressure		
°F	°C	psia	Microns of mercury	Units of vacuum (in Hg)
212	100	14.7	—	0
79	26	0.5	25,400	28.9
72	22	0.4	20,080	29.1*
32	0	0.09	4,579	29.7*
−25	−31	0.005	250	29.8*
−40	−40	0.002	97	29.9*
−60	−51	0.0005	25	29.92*

*Too small a change to be seen on a gauge manifold set.

Step 2. Prior to evacuating a system all of the refrigerant should be recovered.

    A. Refer to Laboratory Worksheets R2, R3, and R4 for recovery procedures.

    B. Also, new systems will need to be evacuated prior to charging.

    C. The evacuation process can also be an indicator for any leaks in the system prior to charging.

Step 3. Connect the vacuum pump, gauge manifold, and the micron gauge as shown in the illustration in Figure R-5-2.

    A. The center hose is connected to a vacuum manifold assembly. This is simply a three valve operation for attaching the vacuum pump and a separate micron gauge, each with a shut-off valve.

    B. Prior to the evacuation, you can check the operation of the vacuum pump by pulling a vacuum on the micron gauge only as shown below.

    C. If the vacuum reaches 200 microns or lower, then the vacuum pump is operating properly.

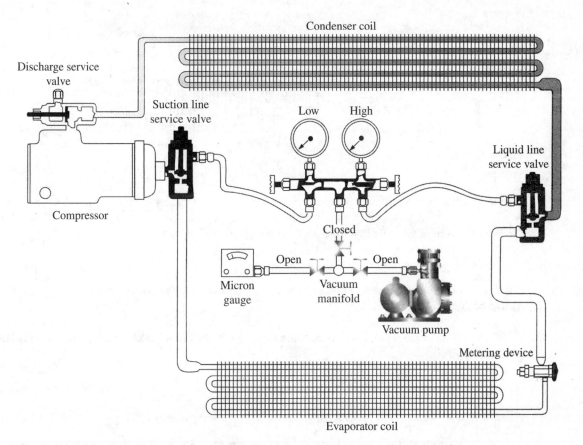

Figure R-5-2

Step 4. Begin evacuation as follows:

    A. Open the valves to the vacuum pump making sure to follow the pump manufacturer's instructions for the pump suction line size, oil, and calibration.

    B. Open wide both valves on the gauge manifold and mid seat both equipment service valves as shown in the illustration in Figure R-5-3.

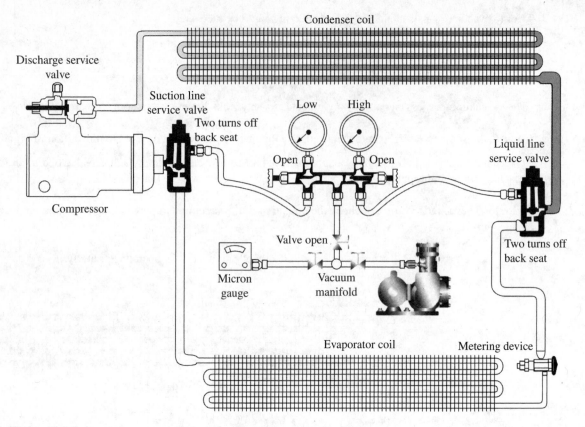

Figure R-5-3

C. Start the vacuum pump and evacuate the system until a vacuum of at least 500 microns is achieved.

D. Close the vacuum pump isolation valve and shut the vacuum pump off.

STEP 5. The micron gauge should still be exposed to the system.

A. If the system maintains under 500 microns for a 10 minute vacuum pressure drop test, the system is considered to be dry and leak free.

B. If 800 to 1,200 microns is maintained, moisture is present in the oil. A leak will cause the micron gauge to rise steadily.

C. A large leak will cause a rapid rise while a small leak will cause a slow rise.

D. Free water in the system will cause a rise to about 20,000 microns.

QUESTIONS

To get help in answering some of the following questions, refer to the *Fundamentals of HVAC/R* text Unit 27.

(Circle the letter that indicates the correct answer.)

1. Vacuum pumps are rated according to:
A. amperage.
B. horsepower.

C. free air displacement (cfm).
D. refrigerant type.

2. One micron is equivalent to:
   A. 1/25,400 of an inch of mercury.
   B. 1/25,400 of an inch of water.
   C. 0.0005 psia.
   D. All of the above are correct.

3. A short evacuation hose with a larger interior diameter:
   A. is better for removing water.
   B. speeds up the evacuation time.
   C. obstructs the flow from the system.
   D. cannot be used for evacuation purposes.

4. Proper evacuation will remove:
   A. air.
   B. water vapor.
   C. noncondensable gases.
   D. All of the above are correct.

5. A deep vacuum is considered to be:
   A. 500 microns or less.
   B. 500 microns or more.
   C. less than 200 microns.
   D. slightly more than 200 microns.

6. The main reason for a deep vacuum is to:
   A. suck out any refrigerant oil in the system.
   B. suck out any water in the system.
   C. Both A and B are correct.
   D. allow any water to flash to vapor so that it can be removed.

7. A system is considered to be dry and leak free if it can maintain a vacuum of less than 500 microns with the pump shut off for a period of:
   A. twenty-four hours.
   B. 10 min.
   C. three hours and 15 min.
   D. about one half hour.

8. A reading of 0 microns would be:
   A. atmospheric pressure.
   B. 0 psig.
   C. a perfect vacuum.
   D. All of the above are correct.

9. At a pressure of 250 microns, water will boil at a temperature of:
   A. 212°F.
   B. 100°C.
   C. 0°F.
   D. –25°F.

10. At a pressure of 25,400 microns, water will boil at a temperature of:
    A. 212°F.
    B. 26°C.
    C. 0°F.
    D. –25°F.

11. If the micron gauge begins to rise steadily once the vacuum pump is shut off, this usually indicates:
    A. a leak.
    B. that everything is normal.
    C. that the micron gauge is faulty.
    D. All of the above are correct.

12. Free water in the system will cause a rise to about 20,000 microns:
    A. True.
    B. False.

13. When recovering refrigerant from an appliance containing less than 200 pounds of R-22 refrigerant, the system must be pulled down to:
    A. a pressure of 10 inches of mercury.
    B. a pressure of 0 psig.
    C. a pressure of 0 psia.
    D. a pressure of 10 psig.

14. When recovering refrigerant from an appliance containing *more* than 200 lb of R-22 refrigerant, the system must be pulled down to:
    A. a pressure of 10 inches of mercury.
    B. a pressure of 0 psig.
    C. a pressure of 0 psia.
    D. a pressure of 10 psig.

15. When recovering refrigerant from very high pressure equipment, the system must be pulled down to:
    A. a pressure of 10 inches of mercury.
    B. a pressure of 0 psig.
    C. a pressure of 0 psia.
    D. a pressure of 10 psig.

TRIPLE EVACUATION

LABORATORY OBJECTIVE
The student will demonstrate the correct procedure for the triple evacuation method of a refrigeration system.

LABORATORY NOTES
This lab is intended to help students practice the evacuation of a refrigeration system. Proper evacuation of a system will remove noncondensable gases (air, water vapor, and inert gases) and ensure a tight, dry system before charging.

The triple evacuation procedure consists of three consecutive evacuations spaced by two dilutions of a dry gas. Nitrogen is preferred but helium and CO_2 can also be used. The clean dry gas will act as a carrier, mixing with system contamination (air and water) and carrying it out with the subsequent evacuation. It is a time consuming procedure but effective in obtaining a clean dry system.

FUNDAMENTALS OF HVAC/R TEXT REFERENCE
Unit 27

Required Tools and Equipment

Gloves and goggles	Unit 3
Vacuum pump	Unit 27
Service valve wrench	Unit 9
Gauge manifold	Unit 12
Cylinder of nitrogen gas with regulator	
Refrigeration system	

SAFETY REQUIREMENTS
Wear safety goggles and gloves when working on refrigeration systems. Oil in the system can become acidic over time and can cause acid burns to the skin and eyes.

PROCEDURE

STEP 1. In this laboratory worksheet, you will be using a vacuum pump that does not pull a deep vacuum in microns. Therefore the time required for evacuation is critical.

    A. The assumption is made that 1 hr of evacuation will remove system contamination without actually measuring the degree of remaining contamination.

    B. The time is sometimes varied to fit the time available on site. An evacuation of one half hour might be good enough if the system is known to be clean. One hour or longer would be even better, no matter what the system problems.

    C. In moisture removal processes overnight evacuations are common.

STEP 2. Prior to evacuating a system all of the refrigerant should be recovered. Refer to Laboratory Worksheets R2, R3, and R4 for recovery procedures. Also, new systems will need to be evacuated prior to charging. The evacuation process can also be an indicator for any leaks in the system prior to charging.

STEP 3. When evacuating any system using Schrader valves, the cores are sometimes removed. The purpose is to get a better quality evacuation faster. If the cores are removed, they must be put back in. Some technicians feel that the time and effort required to put the cores back in and the danger of contamination back into the system make core removal an ineffective practice.

STEP 4. The triple evacuation procedure consists of three consecutive evacuations spaced by two dilutions of a dry gas. You will record the time for each evacuation, the dilution time, and the dilution pressure.

STEP 5. Connect the vacuum pump, gauge manifold, and gas cylinder as shown in Figure R-6-1.

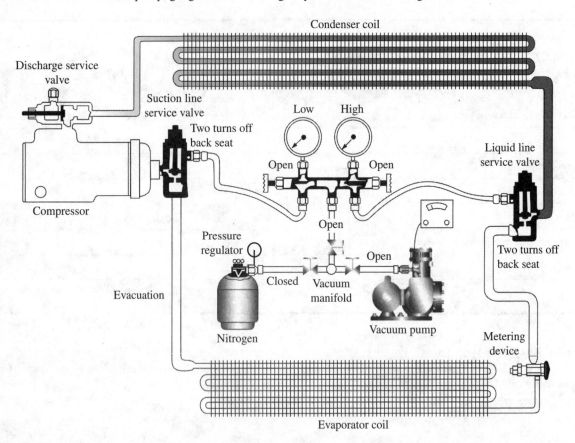

Figure R-6-1

A. The center hose is connected to a vacuum manifold assembly. This is simply a three valve operation for attaching the vacuum pump and the gas cylinder, each with a shut-off valve.

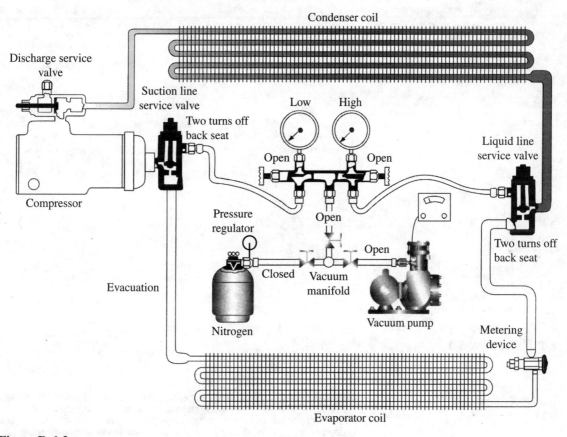

Figure R-6-2

STEP 6. Begin the first evacuation as follows:

A. Both the suction line service valve and the liquid line service valve should be placed in the mid position.

B. Both gauge manifold valves should be wide open.

C. The valves connecting the vacuum pump to the system should be wide open.

D. Start the vacuum pump and run it until the low side compound gauge reads close to 30 inches of vacuum.

E. Close the valve leading to the vacuum pump on the vacuum manifold and shut the vacuum pump off.

F. Record the length of evacuation time.

	Evacuation time	Dilution pressure	Dilution time
1st			
2nd			
3rd			

STEP 7. Open the nitrogen cylinder and adjust the pressure regulator as follows:

A. Make sure that the nitrogen cylinder pressure regulator is turned all the way out (counterclockwise).

B. Slowly open the cylinder valve fully open to back seat it. The tank pressure should register on the regulator high pressure gauge. The pressure in the tank can be in excess of 2,000 psi. *Do not stand in front of the regulator T handle.*

C. Slowly turn the regulator "tee" handle inward (clockwise) until the regulator adjusted pressure reaches approximately 50 psig.

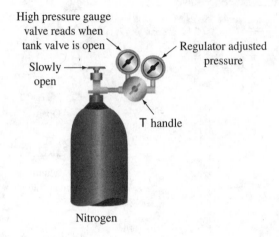

Figure R-6-3

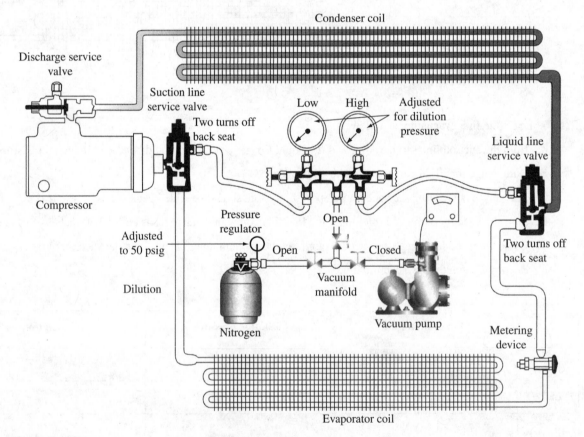

Figure R-6-4

144

STEP 8. Begin the first dilution as follows:

A. Close both valves on the gauge manifold.

B. Open the vacuum manifold isolation valve leading to the gauge manifold and purge any air in the hose between the cylinder pressure regulator and the gauge manifold.

C. Open both gauge manifold valves slowly and obtain the desired dilution pressure (generally 10 psig).

STEP 9.

A. After the dilution period is complete (generally 10 min), close the shut-off valve on the nitrogen cylinder.

B. Disconnect the hose from the vacuum pump and open the closed valve on the vacuum manifold to bleed the dilution pressure to 0 psig.

C. Once the pressure has been bled, you may back all the way off on the T handle (counterclockwise) for the pressure regulator.

STEP 10.

A. Repeat the process over again starting with Step 5 by performing an evacuation.

B. Follow up with a second dilution and then another final evacuation.

QUESTIONS

To get help in answering some of the following questions, refer to the *Fundamentals of HVAC/R* text Unit 27.

(Circle the letter that indicates the correct answer.)

1. Triple evacuation refers to:
   A. using three separate vacuum pumps.
   B. evacuating a refrigeration system three consecutive times.
   C. an evacuation level that increases by thirds.
   D. an evacuation level that decreases by thirds.

2. Dry gases that can be used in a triple evacuation process include:
   A. nitrogen.
   B. helium.
   C. CO_2.
   D. All of the above are correct.

3. Nitrogen cylinder pressures:
   A. are never above 150 psig.
   B. are generally between 350 to 400 psig.
   C. can be as high as 2,000 psig.
   D. can sometimes be negative (vacuum).

4. Proper evacuation will remove:
   A. air.
   B. refrigerant.
   C. oil.
   D. All of the above are correct.

5. A triple evacuation may be performed:
   A. because it is very fast.
   B. only on systems containing CFC refrigerants.

C. if a deep vacuum pump is unavailable.
D. on ammonia systems only.

6. The reason for introducing a clean dry gas between evacuations is to:
    A. mix with the system contaminants.
    B. destroy the system contaminants.
    C. ensure a tight dry system.
    D. check for leaks.

7. Nitrogen cylinder tank valves should be:
    A. never opened fully.
    B. opened fully to back seat them.
    C. partially throttled.
    D. normally closed as there is a relief hole drilled through the seat of the valve.

8. A number of different dry gases may be used for a triple evacuation such as:
    A. nitrogen, helium, and CO_2.
    B. nitrogen, helium, and O_2.
    C. nitrogen, helium, and H_2O.
    D. All of the above are correct.

9. Nitrogen cylinder tank pressures are relatively low:
    A. True.
    B. False.

VACUUM PRESSURE DROP TEST

LABORATORY OBJECTIVE

The student will demonstrate the correct procedure for the vacuum pressure drop test to check for leaks in a refrigeration system.

LABORATORY NOTES

To prepare for this lab, students must complete the triple evacuation procedure as outlined in Laboratory Worksheet R-6.

FUNDAMENTALS OF HVAC/R TEXT REFERENCE

Unit 27

Required Tools and Equipment

Gloves and goggles	Unit 3
Vacuum pump	Unit 27
Service valve wrench	Unit 9
Gauge manifold	Unit 12
Cylinder of inert gas	
Refrigeration system	

SAFETY REQUIREMENTS

Wear safety goggles and gloves when working on refrigeration systems. Oil in the system can become acidic over time and can cause acid burns to the skin and eyes.

PROCEDURE

STEP 1. The timed vacuum pressure drop test is typically done at the conclusion of a timed evacuation or a triple evacuation (Laboratory Worksheet R-6) and used as an additional leak testing procedure.

A. If the system holds approximately 30 inches of vacuum overnight or over a weekend, there can be no leaks. This is a valid procedure and should be done as time permits at the discretion of the service technician and company policy.

B. Remember that this procedure will only help determine the existence of a leak in the system and does not pinpoint the location of the leak.

STEP 2. Upon the final evacuation of a triple evacuation procedure or at the end of a timed evacuation, close both gauge manifold valves with the vacuum pump still operating and then shut off the pump.

STEP 3. The unit will be periodically checked to determine if the system is tight and holding a vacuum for a specified time period.

QUESTIONS

To get help in answering some of the following questions, refer to the *Fundamentals of HVAC/R* text Unit 27.

(Circle the letter that indicates the correct answer.)

1. A vacuum pressure drop test:
   A. will be able to pinpoint the location of a leak.
   B. will be able to provide an approximate location of a leak.
   C. indicates that there is a leak somewhere.
   D. will verify the accuracy of the vacuum pump.

2. A vacuum leak can be detected by:
   A. an electronic leak detector.
   B. a halide torch.
   C. bubble test.
   D. an ultrasonic type leak detector.

3. A triple evacuation:
   A. is normally performed with a deep vacuum pump.
   B. will evacuate the system three times.
   C. can be done with refrigerant still remaining in the system.
   D. will verify the accuracy of the gauge manifold.

4. When testing for a leak with an electronic leak detector:
   A. remember that most refrigerants are heavier than air and will settle to the bottom.
   B. remember that most refrigerants are lighter than air and will rise to the top.
   C. move the tip of the detector across the suspected leak area very quickly.
   D. always test the outlet of the vacuum pump first.

5. High vacuum indicators have accuracy approaching:
   A. 500 microns.
   B. 50 microns.
   C. 10 microns.
   D. 2 microns.

6. When a system is evacuated:
   A. water in the system will vaporize.
   B. water vapor in the system will condense.
   C. Both A and B are correct.
   D. None of the above is correct.

7. Most compound gauges:
   A. can read pressure and vacuum.
   B. can read vacuum in microns.

C. Both A and B are correct.
D. None of the above is correct.

8. If the Schrader valve core is pulled from a service port:
   A. a vacuum can be pulled much quicker.
   B. this should be done without opening the system to the atmosphere.
   C. Both A and B are correct.
   D. None of the above is correct.

NITROGEN PRESSURE TEST

LABORATORY OBJECTIVE

The student will demonstrate the correct procedure for a nitrogen pressure test to check for leaks in a refrigeration system.

LABORATORY NOTES

After a system has been repaired and before the final charge has been installed, the system needs to be leak tested. This is done by pressurizing the system with an inert gas such as nitrogen or carbon dioxide mixed with a trace amount of 5% or 10% R-22.

Oxygen should not be used, since it can cause an explosion.

Under current EPA regulations, the mixture of R-22 and inert gas is not considered to be refrigerant. Therefore it can be vented to the atmosphere following the leak check.

However, the EPA does not look favorably on the addition of the inert gas to the refrigerant already present in a system. This is the reason why you must completely evacuate the remaining refrigerant in the system before introducing the trace gas.

This lab is intended to allow student practice on basic leak detection procedures. Since it is not the intention to recharge the system at the completion of this lab, an unrepaired leak could be set up and left for students to find. On such a system it would be impossible to pass a micron evacuation or any vacuum pressure drop test.

FUNDAMENTALS OF HVAC/R TEXT REFERENCE
Unit 26

Required Tools and Equipment

Gloves and goggles	Unit 3
Service valve wrench	Unit 9
Gauge manifold	Unit 12
Electronic leak detector/halide torch	Unit 26
Cylinder of nitrogen gas with regulator	
Cylinder of refrigerant R-22	
Refrigeration system that has a leak	

SAFETY REQUIREMENTS

A. Wear safety goggles and gloves when working with refrigerants. Liquid refrigerant can cause frostbite when in contact with eyes and skin.

B. Use low loss hose fittings, or wrap cloth around hose fittings before removing the fittings from a pressurized system or cylinder. Inspect all fittings before attaching hoses.

PROCEDURE

STEP 1. Prior to performing this procedure, all of the refrigerant should be recovered. Refer to Laboratory Worksheets R2, R3, and R4 for recovery procedures.

STEP 2. Connect the R-22 refrigerant cylinder, gauge manifold, and the gas cylinder as shown in Figure R-8-1.

A. The center hose is connected to a vacuum manifold assembly. This is simply a three valve operation for attaching the refrigerant cylinder and the gas cylinder, each with a shut-off valve. If you do not have this type of arrangement, you can connect each item separately.

B. Purge any air from the hose connection to the gauge manifold prior to introducing the refrigerant into the system.

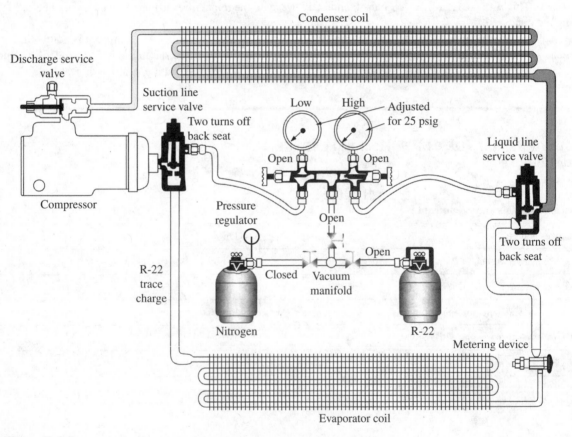

Figure R-8-1

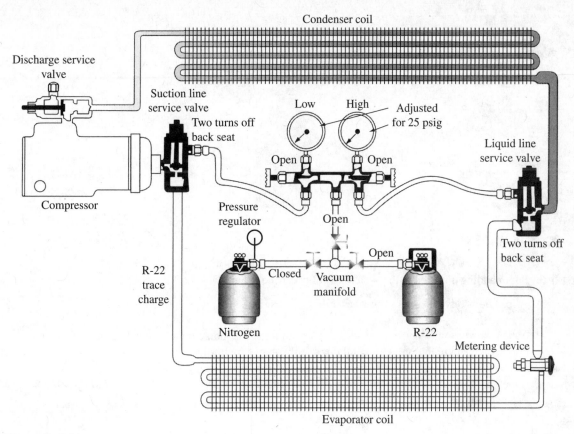

Figure R-8-2

STEP 3. Admit the trace charge as follows:

A. After the lines have been purged, slowly admit R-22 vapor to the system through both the high side and low side valves as a trace charge of 25 psig.

B. When the trace charge has been added, close the cylinder, the gauge manifold valves, and the vacuum manifold valve.

C. You are only adding a small amount of refrigerant so that it can be detected by the leak detector.

D. If only pure nitrogen were added, it would be difficult to find the leak. With pure nitrogen only, you would need to use the soap and bubble method or an ultrasonic leak detector.

STEP 4. Open the nitrogen cylinder and adjust the pressure regulator as follows:

A. Make sure that the nitrogen cylinder pressure regulator is turned all the way out (counterclockwise).

B. Slowly open the cylinder valve fully open to back seat it. The tank pressure should register on the regulator high pressure gauge. The pressure in the tank can be in excess of 2,000 psi. *Do not stand in front of the regulator T handle.*

C. Slowly turn the regulator T handle inward (clockwise) until the regulator adjusted pressure reaches approximately 150 psig.

153

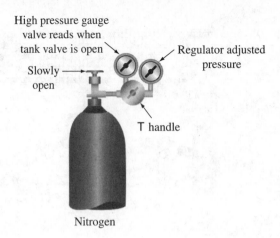

High pressure gauge
valve reads when
tank valve is open

Regulator adjusted
pressure

Slowly
open

T handle

Nitrogen

Figure R-8-3

STEP 5. Begin adding nitrogen to the system as follows (See Figure R-8-4):

A. Close both valves on the gauge manifold.

B. Open the vacuum manifold isolation valve leading to the gauge manifold and purge any air in the hose between the cylinder pressure regulator and the gauge manifold.

C. Open both gauge manifold valves slowly and obtain the desired nitrogen pressure of approximately 150 psig. Then close the gauge manifold valves.

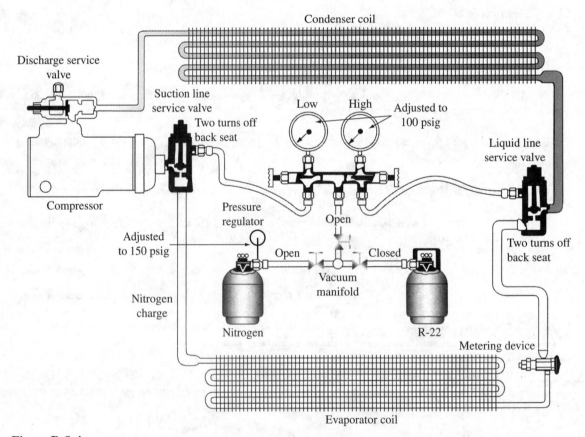

Figure R-8-4

154

STEP 6.

A. With the system pressure now at 150 psig, you may begin checking for leaks using an electronic leak detector or halide torch.

B. Record all leak locations and consult with your instructor before making any repairs.

STEP 7.

A. After all of the leaks have been located, you may drain the system.

B. Close the shut-off valve on the nitrogen cylinder. Carefully disconnect the hose from the R-22 refrigerant cylinder and open the closed valve on the vacuum manifold to bleed the system pressure to 0 psig.

C. Once the pressure has been bled, you may back all the way off on the T handle (counterclockwise) for the pressure regulator on the nitrogen cylinder.

QUESTIONS

To get help in answering some of the following questions, refer to the *Fundamentals of HVAC/R* text Unit 26.

(Circle the letter that indicates the correct answer.)

1. According to EPA regulations:
   A. nitrogen can be added to a charged system for leak testing.
   B. pure CFCs or HCFs can be released for leak testing.
   C. nitrogen mixed with 5% or 10% R-22 can be used for leak testing.
   D. All of the above are correct.

2. To check for refrigerant leaks you may use:
   A. a halide leak detector.
   B. an electronic leak detector.
   C. a bubble test.
   D. All of the above are correct.

3. If you use 100% nitrogen for leak testing:
   A. a halide leak detector must be used.
   B. a bubble test must be performed.
   C. an electronic leak detector may be used.
   D. All of the above are correct.

4. An ultrasonic type leak detector:
   A. can detect any gas leaking through an orifice.
   B. detects pressure or vacuum leaks.
   C. detects ultrasonic noise from arcing electrical switchgear.
   D. All of the above are correct.

5. An electronic leak detector works by:
   A. drawing air over a platinum diode.
   B. a magnetic impulse.
   C. sending out an electric charge.
   D. None of the above is correct.

VAPOR CHARGING WITH CHARGING CYLINDER

LABORATORY OBJECTIVE

The student will demonstrate the correct procedure for performing a vapor charge using a charging cylinder on a small refrigeration unit.

LABORATORY NOTES

Prior to charging, the system must be leak tested and evacuated (refer to Laboratory Worksheets R-5 and R-6 for the correct procedures for evacuation). In this way, when the charging is started, the system is under vacuum so that when the refrigerant enters the system it will be drawn into the unit due to the difference in pressure.

When charging refrigerants that fractionate (zeotropes), such as R-401a, they must be charged as a liquid to prevent separation of the refrigerant as it enters the system.

FUNDAMENTALS OF HVAC/R **TEXT REFERENCE**
Unit 27

Required Tools and Equipment

Gloves and goggles	Unit 3
Service valve wrench	Unit 9
Gauge manifold	Unit 12
Charging cylinder	Unit 27
Cylinder of refrigerant	
Operating refrigeration system	

SAFETY REQUIREMENTS

    A. Wear safety goggles and gloves when working with refrigerants. Liquid refrigerant can cause frostbite when in contact with eyes and skin.

    B. Use low loss hose fittings, or wrap cloth around hose fittings before removing the fittings from a pressurized system or cylinder. Inspect all fittings before attaching hoses.

PROCEDURE

STEP 1. The refrigerant type and required amount of charge can be found on the refrigeration system name plate as shown in Figure R-9-1.

F.L.A.		F.L.A.	
L.R.A.		L.R.A.	
H.P.		H.P.	
Volts		Volts	
Hertz phase		Hertz phase	

R-22 | 5.0 | lb | | kg

Figure R-9-1

STEP 2.

A. Connect a charging cylinder, a vacuum pump, and the refrigerant cylinder with a gauge manifold as shown in Figure R-9-2.

B. Start the vacuum pump to draw air from the charging cylinder and the hoses. The vacuum in the charging cylinder will also allow for a better flow from the refrigerant cylinder to the charging cylinder.

C. When vacuum has been achieved in the charging cylinder, close the low side gauge manifold valve and secure the vacuum pump and remove it.

STEP 3. You must be able to accurately read the refrigerant level on a charging cylinder.

A. Refrigerant is always measured by weight in pounds and ounces or kilograms and grams in SI units. The measured volume on the charging cylinder is calibrated to an equivalent weight on the sliding scale. See Figure R-9-3.

B. Since the temperature and pressure of the refrigerant will have a direct effect on its volume, a sliding weight scale is used. Some charging cylinders have scales for more than one type of refrigerant, so make sure that you are using the proper scale.

C. The scale should be turned to adjust for the pressure registered by the pressure gauge on the top of the charging cylinder.

STEP 4.

A. Close the high side gauge manifold valve as shown in Figure R-9-4.

B. Open the valve on the refrigerant cylinder and then turn it upside down.

C. Slowly crack and throttle the high side gauge manifold valve to begin filling the charging cylinder with the correct amount of liquid refrigerant.

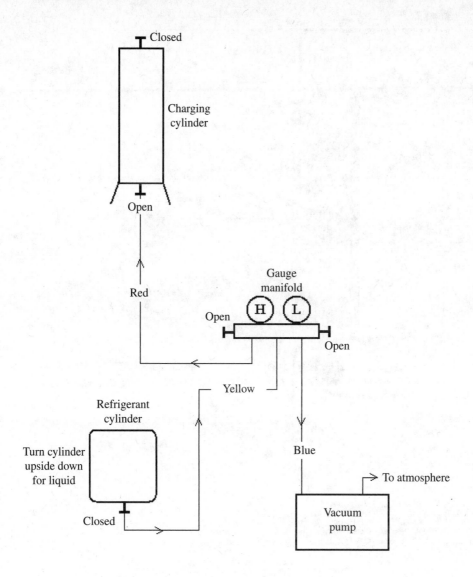

Figure R-9-2

D. Read the weight registered on the scale for the volume of refrigerant that you have added to the cylinder and record it here.

_____ lb _____ oz

STEP 5.

A. When the charging cylinder is full with the proper amount of refrigerant, close the filling valve on the bottom of the charging cylinder and also the valve on the refrigerant cylinder.

B. Drain any excess pressure in the hoses through the gauge manifold low pressure hose and then carefully disconnect the charging cylinder and refrigerant cylinder from the gauge manifold.

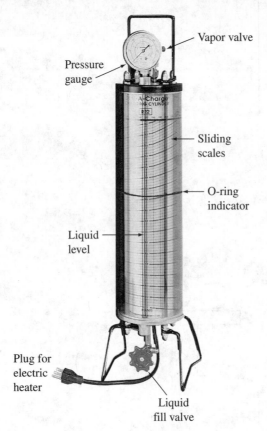

Pressure gauge

Vapor valve

Sliding scales

O-ring indicator

Liquid level

Plug for electric heater

Liquid fill valve

Figure R-9-3

STEP 6. You should have already determined the amount of charge required for the system.

A. Subtract the amount of charge required from the amount measured in the cylinder and you will be able to determine when the charge is complete.

B. As an example, assume you have filled the cylinder to a level of 22 oz after dialing the scale to match the appropriate pressure. Also assume the required charge for the refrigeration system is 16 oz.

You would charge until the charging cylinder reached a level of 22 − 16 = 6 oz.

C. There is an O-ring indicator that can be moved up and down the scale. You would slide the O-ring to the level of 6 oz. This will help you remember when to stop the charge.

D. Many charging cylinders also have electric heaters built into them. They can be plugged into a regular electrical wall outlet. This will raise the pressure in the charging cylinder and speed up the charging process.

E. If the charging cylinder does not have an electrical heater, you may try to elevate the cylinder pressure by placing it in a bucket of warm water.

NEVER use an open flame to heat a refrigerant cylinder!!

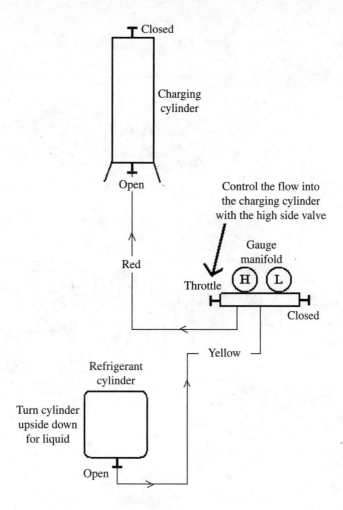

Figure R-9-4

STEP 7. Begin charging as follows:

A. Connect the gauge manifold to the vapor side of the charging cylinder as shown in Figure R-9-5.

B. Purge the air from the lines while connecting the remaining hoses on the gauge manifold to the high and low side of the refrigeration system that you are charging (Refer to Laboratory Worksheet R-1 *Basic Refrigeration System Startup,* Step 4).

C. If the system has Schrader valves, then be careful not to allow air to enter the system when you connect the hoses.

D. The system should already have been evacuated and in a vacuum and the refrigerant should flow freely as a vapor from the top of the charging cylinder into both the high side and low side of the system as a vapor.

E. Use the electric heater on the charging cylinder to the increase the speed of the charge.

STEP 8. At some point the pressure in the refrigeration system will equalize with the pressure in the charging cylinder and no more refrigerant will flow. This may happen even though the full charge is not yet complete. At this point you will need to run the system to draw the remaining refrigerant from the charging cylinder into the low side.

A. To do this, first close the high side valve on the gauge manifold before running the system.

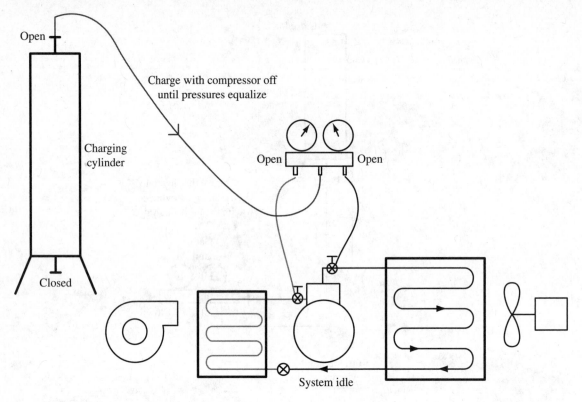

Open

Charge with compressor off
until pressures equalize

Charging
cylinder

Open

Open

Closed

System idle

Figure R-9-5

B. When the system is turned on, the refrigerant should once again begin to flow, this time into the low side (suction) only.

C. Although there is no flow through the high side of the gauge manifold, you will still be able to monitor the system discharge pressure on the high side gauge.

D. When the proper amount of refrigerant has been added to the system, close the gauge manifold low side valve.

E. Even though the high side and low side gauge manifold valves are now both closed, you will still be able to monitor the refrigeration system pressures due to the configuration of the gauge manifold.

F. Keep the gauge manifold connected long enough to allow the system time to stabilize.

G. Verify that the suction and discharge pressures of the refrigeration system agree with the manufacturer's recommendations and record the pressures below.

Suction pressure _____ Discharge pressure _____

H. After the system has been charged and the operating pressures verified, back seat the suction and discharge service valves and close the vapor valve on the charging cylinder.

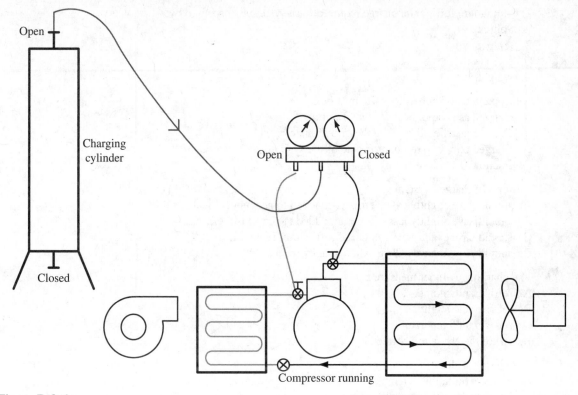

Figure R-9-6

I. Carefully disconnect the hoses.

J. Any excess refrigerant remaining in the charging cylinder can be used for subsequent charges or can be transferred to a recovery cylinder through a recovery unit.

QUESTIONS

To get help in answering some of the following questions, refer to the *Fundamentals of HVAC/R* text Unit 27.

(Circle the letter that indicates the correct answer.)

1. It is usually considered good practice to charge with vapor:
   A. to prevent any danger of slugging the compressor with liquid.
   B. because it is faster than a liquid charge.
   C. because zeotropes must be charged as vapors.
   D. to ensure accurate measurement of the charge.

2. A color coded disposable refrigeration cylinder in the upright position:
   A. has vapor at the top and liquid on the bottom.
   B. has liquid at the top and vapor at the bottom.
   C. would release liquid if the cylinder valve were opened.
   D. would release a mixture of liquid and vapor if the cylinder valve were opened.

3. The amount of refrigerant charge required for a system is measured by:
   A. volume.
   B. temperature.
   C. pressure.
   D. weight.

4. Zeotropic blends such as R-401a must be:
   A. charged as a vapor because they fractionate.
   B. charged as a liquid because they fractionate.
   C. charged only into the suction side of the system.
   D. fractionated before charging.

5. A properly charged system:
   A. should have slightly extra refrigerant added to account for leaks.
   B. should have slightly less refrigerant added to prevent freeze-up.
   C. should have exactly the amount of refrigerant required.
   D. should always have a 20% extra capacity reserve.

6. The charging cylinder measures:
   A. refrigerant volume.
   B. refrigerant weight.
   C. refrigerant temperature.
   D. All of the above are correct.

7. The sliding scale on a charging cylinder:
   A. corrects for ambient temperature.
   B. corrects for ambient pressure.
   C. is seldom of any use.
   D. serves as a refrigerant saturation table.

8. Charging cylinders may have built in electric heaters:
   A. so they do not freeze.
   B. to drive off any moisture mixed with the refrigerant.
   C. to reduce fractionation.
   D. to speed up the rate of charge.

9. A liquid charge is always much faster than a vapor charge:
   A. True.
   B. False.

VAPOR CHARGING WITH DIGITAL SCALE

LABORATORY OBJECTIVE

The student will demonstrate the correct procedure for performing a vapor charge on a small refrigeration unit using a digital scale.

LABORATORY NOTES

Prior to charging, the system must be leak tested and evacuated (refer to Laboratory Worksheets R-5 and R-6 for the correct procedures for evacuation). In this way, when the charging is started, the system is under vacuum so that when the refrigerant enters the system it will be drawn into the unit due to the difference in pressure.

When charging refrigerants that fractionate (zeotropes), such as R-401a, they must be charged as a liquid to prevent separation of the refrigerant as it enters the system.

FUNDAMENTALS OF HVAC/R TEXT REFERENCE
Unit 27

Required Tools and Equipment

Gloves and goggles	Unit 3
Service valve wrench	Unit 9
Gauge manifold	Unit 12
Digital scale	Unit 27
Cylinder of refrigerant	
Operating refrigeration system	

SAFETY REQUIREMENTS

A. Wear safety goggles and gloves when working with refrigerants. Liquid refrigerant can cause frostbite when in contact with eyes and skin.

B. Use low loss hose fittings, or wrap cloth around hose fittings before removing the fittings from a pressurized system or cylinder. Inspect all fittings before attaching hoses.

PROCEDURE

STEP 1. The refrigerant type and required amount of charge can be found on the refrigeration system name plate as shown in Figure R-10-1.

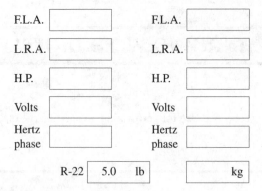

F.L.A.			F.L.A.	
L.R.A.			L.R.A.	
H.P.			H.P.	
Volts			Volts	
Hertz phase			Hertz phase	
	R-22	5.0 lb		kg

Figure R-10-1

STEP 2. Begin charging as follows:

A. Place a refrigerant cylinder on a digital scale in the upright position and connect the gauge manifold as shown in Figure R-10-2.

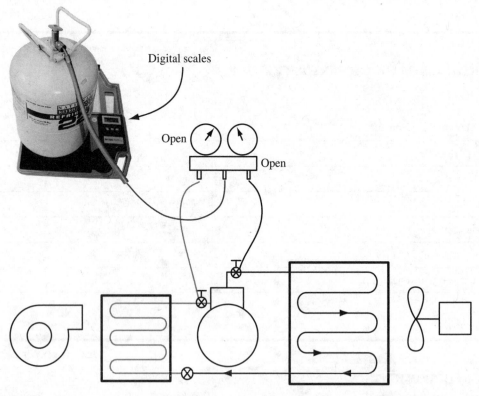

Digital scales

Open

Open

Figure R-10-2

B. Purge the air from the lines while connecting the remaining hoses from the gauge manifold to the high and low side of the refrigeration system that you are charging (Refer to Laboratory Worksheet R-1 *Basic Refrigeration System Startup,* Step 4).

C. If the system has Schrader valves, then be careful not to allow air to enter the system when you connect the hoses.

D. Set the digital readout on the scale to zero.

E. After the lines have been purged, you are ready to begin charging. First the service valves on the refrigeration system can be opened, then the gauge manifold valves can be opened, and then the refrigerant cylinder can be opened.

F. The system should already have been evacuated and in a vacuum. The refrigerant should flow freely into both the high side and low side of the system as a vapor.

G. Watch the digital readout as you charge and the weight of the cylinder should be decreasing by a negative amount. As an example, assume the system requires a 12 oz charge. You would continue to charge until the readout measured –12 oz (minus 12 oz).

H. It is unlikely that you will overcharge the system at this point because normally only 50% to 75% of the charge will flow before the charging stops and the pressures equalize.

I. You can try to speed up the rate of charge by placing the cylinder in a bucket of warm water. If you choose to do this, then the bucket should be in place before you set the digital scale reading to zero or the charge quantity will be incorrect. Digital scales are very sensitive and the slightest movement can alter the reading.

NEVER let the temperature of the water bath exceed 125°F.

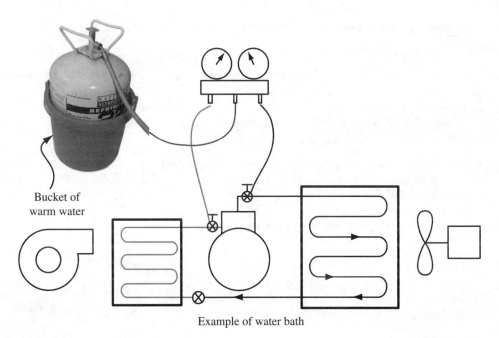

Bucket of
warm water

Example of water bath

Figure R-10-3

STEP 3. At some point, the pressure in the refrigeration system will equalize with the pressure in the charging cylinder and no more refrigerant will flow. This may happen even though the full charge is not yet complete. At this point you will need to run the system to draw the remaining refrigerant from the refrigerant cylinder into the low side.

A. To do this, first close the high side valve on the gauge manifold before running the system.

B. When the system is turned on, the refrigerant should once again begin to flow, this time into the low side (suction) only.

C. Although there is no flow through the high side of the gauge manifold, you will still be able to monitor the system discharge pressure on the high side gauge.

D. When the proper amount of refrigerant has been added to the system, close the gauge manifold low side valve.

E. Even though the high side and low side gauge manifold valves are now both closed, you will still be able to monitor the refrigeration system pressures due to the configuration of the gauge manifold.

F. Keep the gauge manifold connected long enough to allow the system time to stabilize.

G. Verify that the suction and discharge pressures of the refrigeration system agree with the manufacturer's recommendations and record the pressures below.

Suction pressure _____ Discharge pressure _____

H. After the system has been charged and the operating pressures verified, back seat the suction and discharge service valves and close the valve on the refrigerant cylinder.

I. Carefully disconnect the hoses.

QUESTIONS

To get help in answering some of the following questions, refer to the *Fundamentals of HVAC/R* text Unit 27.

(Circle the letter that indicates the correct answer.)

1. When using a digital scale it is good practice to place it on a sturdy and level platform because:
   A. any slight movement can alter the readings.
   B. if the refrigerant cylinder is stable, you will get a more accurate reading.
   C. the digital scale is very accurate and therefore somewhat sensitive.
   D. All of the above are correct.

2. When using a warm water bath to heat a refrigeration cylinder:
   A. make sure that distilled water is used.
   B. never exceed 125°F.
   C. always use a porcelain container.
   D. temperatures as high as 250°F are often obtained.

3. A digital scale would be:
   A. much faster to use than a charging cylinder.
   B. much slower to use than a charging cylinder.
   C. used to measure the weight of a charging cylinder.
   D. used to measure the weight of the gauge manifold.

4. Prior to charging:
   A. the system should have a press charge of nitrogen added.
   B. the system should be flushed with air.
   C. the system must be leak tested and evacuated.
   D. the system must be vented to the atmosphere.

5. Charging with a vapor:
   A. is cleaner than charging with a liquid.
   B. is not as clean as charging with a liquid.
   C. is faster than charging with a liquid.
   D. may damage the compressor due to a liquid slug.

6. The digital scale is set to zero:
   A. allowing it to calibrate.
   B. so that an accurate measurement of refrigerant can be obtained.
   C. halfway through the charging process.
   D. three quarters of the way through the charging process.

7. An insufficient charge could lead to:
   A. flooding of the evaporator.
   B. flooding of the compressor.
   C. flooding of the condenser.
   D. starving of the evaporator.

8. About a 1% change in refrigerant charge will change the superheat:
   A. 5°F or more.
   B. 15°F or more.
   C. 3°F or more.
   D. 25°F or more.

9. A vapor charge is always much faster than a liquid charge:
   A. True.
   B. False.

LIQUID CHARGING WITH COMPRESSOR OFF

LABORATORY OBJECTIVE

The student will demonstrate the correct procedure for performing a liquid charge on a small refrigeration unit using a digital scale.

LABORATORY NOTES

Prior to charging, the system must be leak tested and evacuated (refer to Laboratory Worksheets R-5 and R-6 for the correct procedures for evacuation). In this way, when the charging is started, the system is under vacuum so that when the refrigerant enters the system it will be drawn into the unit due to the difference in pressure.

When charging refrigerants that fractionate (zeotropes), such as R-401a, they must be charged as a liquid to prevent separation of the refrigerant as it enters the system.

FUNDAMENTALS OF HVAC/R TEXT REFERENCE
Unit 27

Required Tools and Equipment

Gloves and goggles	Unit 3
Service valve wrench	Unit 9
Gauge manifold	Unit 12
Digital scale	Unit 27
Cylinder of refrigerant	
Operating refrigeration system	

SAFETY REQUIREMENTS

A. Wear safety goggles and gloves when working with refrigerants. Liquid refrigerant can cause frostbite when in contact with eyes and skin.

B. Use low loss hose fittings, or wrap cloth around hose fittings before removing the fittings from a pressurized system or cylinder. Inspect all fittings before attaching hoses.

PROCEDURE

STEP 1. The refrigerant type and required amount of charge can be found on the refrigeration system name plate as shown in Figure R-11-1.

F.L.A. [   ]   F.L.A. [   ]

L.R.A. [   ]   L.R.A. [   ]

H.P. [   ]   H.P. [   ]

Volts [   ]   Volts [   ]

Hertz [   ]   Hertz [   ]
phase   phase

R-22 [5.0] lb   [   ] kg

Figure R-11-1

STEP 2.

    A. Liquid charging is always much faster than charging with vapor.

    B. Liquid is charged on the high side of the system.

    C. On small systems the charging is done with the compressor off and the full charge is seldom completed.

    D. Therefore the vapor method must be used to complete the process.

    E. You must also make sure that no liquid enters the compressor.

STEP 3.

    A. Place a refrigerant cylinder on a digital scale in the upright position and connect the gauge manifold as shown in Figure R-11-2.

    B. Purge the air from the lines while connecting the remaining hoses from the gauge manifold to the high and low side of the refrigeration system that you are charging (Refer to Lab R-1 *Basic Refrigeration System Startup,* Step 4).

    C. If the system has Schrader valves, then be careful not to allow air to enter the system when you connect the hoses.

STEP 4. After the lines have been purged you are ready to begin charging as follows:

    A. Normally the refrigerant cylinder must be turned upside down (inverted) for liquid.

    B. It is always good practice to open the cylinder valve all the way open prior to turning it over.

    C. Once the bottle is inverted and balanced on the scale, set the digital readout on the scale to zero.

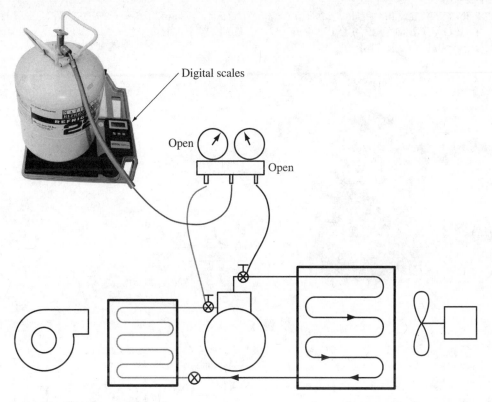

Digital scales

Open

Open

Figure R-11-2

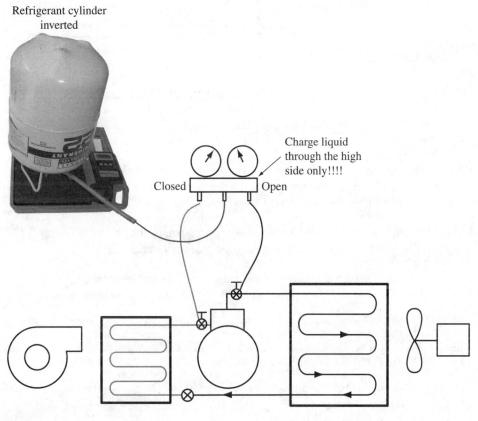

Refrigerant cylinder
inverted

Charge liquid
through the high
side only!!!!

Closed

Open

Figure R-11-3

173

D. Prior to opening the service valves on the refrigeration system, make sure that ONLY THE HIGH SIDE OF THE GAUGE MANIFOLD IS OPEN AND THE LOW SIDE GAUGE MANIFOLD VALVE IS CLOSED as shown in Figure R-11-4.

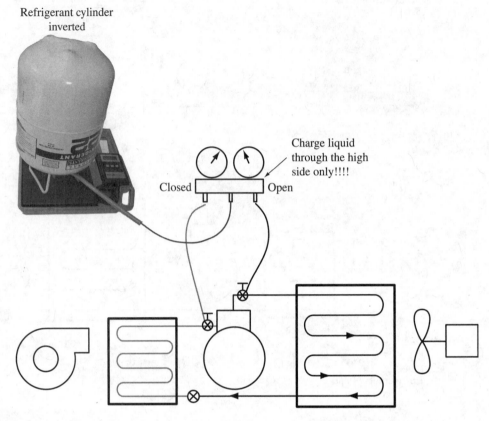

Figure R-11-4

E. NEVER ADMIT LIQUID INTO THE SUCTION SIDE OF THE COMPRESSOR!!

F. The system should already have been evacuated and in a vacuum. When the high side service valve is opened, the refrigerant should flow freely into the high side of the system as a liquid.

G. Watch the digital readout as you charge and the weight of the cylinder should be decreasing by a negative amount. As an example, assume the system requires a 12 oz charge. You would continue to charge until the readout measured –12 oz (minus 12 oz).

H. It is unlikely that you will overcharge the system at this point because normally only 50% to 75% of the charge will flow before the charging stops and the pressures equalize.

I. At this point close the high side valve on the gauge manifold and record the readout in pounds and ounces from the digital scale in the space provided below.

Amount of liquid charge _____lb_____oz_____

STEP 5. If the full charge is not complete after adding the liquid, then the vapor method must be used to complete the process.

A. Calculate the remaining refrigerant required to complete the charge. Subtract the amount of liquid already added from the total charge required.

Total charge required − liquid added = vapor charge remaining

B. Turn the cylinder back over so that it is upright for vapor.

C. Start the system in the normal run mode and the discharge pressure should register on the high side of the gauge manifold.

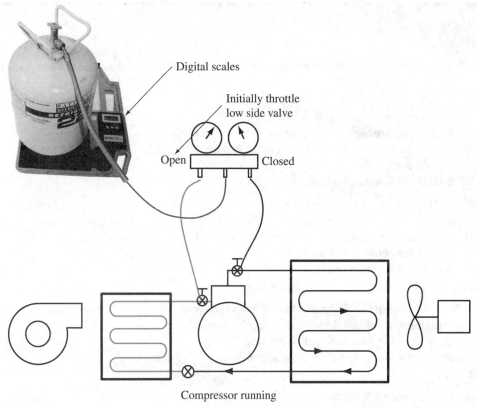

Figure R-11-5

D. Although there is no flow through the high side of the gauge manifold, you will still be able to monitor the system discharge pressure on the high side gauge.

E. Since there has been liquid in the hoses, you will throttle the gauge manifold low side valve as you begin vapor charging. This will allow any remaining liquid left over in the hose to flash so as not to introduce liquid into the suction side of the compressor.

F. Slowly open the refrigeration system suction service valve while throttling the low side gauge manifold valve.

G. When the proper amount of refrigerant has been added to the system, close the gauge manifold low side valve.

H. Even though the high side and low side gauge manifold valves are now both closed, you will still be able to monitor the refrigeration system pressures due to the configuration of the gauge manifold.

I. Keep the gauge manifold connected long enough to allow the system time to stabilize.

J. Verify that the suction and discharge pressures of the refrigeration system agree with the manufacturer's recommendations and record the pressures below.

Suction pressure _____ Discharge pressure _____

K. After the system has been charged and the operating pressures verified, back seat the suction and discharge service valves and close the valve on the refrigerant cylinder.

L. Carefully disconnect the hoses.

QUESTIONS

To get help in answering some of the following questions, refer to Unit 27 of the *Fundamentals of HVAC/R* text.

(Circle the letter that indicates the correct answer.)

1. If the refrigeration system is overcharged:
   A. the discharge pressure will be higher than expected.
   B. the discharge pressure will be lower than expected.
   C. the box temperature will be too warm.
   D. Both B and C are correct.

2. If the refrigeration system is *under*charged:
   A. the discharge pressure will be higher than expected.
   B. the discharge pressure will be lower than expected.
   C. the box temperature will be too warm.
   D. Both B and C are correct.

3. When liquid charging:
   A. never allow liquid refrigerant to enter the compressor.
   B. small amounts of liquid may enter the compressor.
   C. charge into the suction side of the system.
   D. charge into the suction side of the system with a gauge manifold.

4. Throttling liquid refrigerant through the gauge manifold:
   A. will increase its temperature.
   B. will allow some of it to flash to vapor.
   C. will damage the gauge manifold.
   D. will damage the gauge manifold valve seats.

5. To speed up the charging process:
   A. heat the cylinder with a propane torch.
   B. heat the cylinder to at least 150°F.
   C. charge with a liquid.
   D. charge with a vapor.

6. When charging, if the gauge manifold valves are closed (front seated):
   A. no flow will occur through the manifold.
   B. system pressures can still be read.
   C. the system is isolated from the charging cylinder.
   D. All of the above are correct.

7. An insufficient charge could lead to starving of the evaporator:
   A. True.
   B. False.

LIQUID CHARGING WITH COMPRESSOR RUNNING

LABORATORY OBJECTIVE

The student will demonstrate the correct procedure for performing a liquid charge on a refrigeration unit that has a king valve.

LABORATORY NOTES

The refrigeration unit to be charged according to this procedure must have a manual king valve located after the condenser.

Prior to charging, the system must be leak tested and evacuated (refer to Laboratory Worksheets R-5 and R-6 for the correct procedures for evacuation). In this way, when the charging is started, the system is under vacuum so that when the refrigerant enters the system it will be drawn into the unit due to the difference in pressure.

When charging refrigerants that fractionate (zeotropes), such as R-401a, they must be charged as a liquid to prevent separation of the refrigerant as it enters the system.

FUNDAMENTALS OF HVAC/R TEXT REFERENCE
Unit 27

Required Tools and Equipment

Gloves and goggles	Unit 3
Service valve wrench	Unit 9
Gauge manifold	Unit 12
Digital scale	Unit 27
Cylinder of refrigerant	
Operating refrigeration system	

SAFETY REQUIREMENTS

A. Wear safety goggles and gloves when working with refrigerants. Liquid refrigerant can cause frostbite when in contact with eyes and skin.

B. Use low loss hose fittings, or wrap cloth around hose fittings before removing the fittings from a pressurized system or cylinder. Inspect all fittings before attaching hoses.

PROCEDURE

STEP 1. The refrigerant type and required amount of charge can be found on the refrigeration system name plate as shown in Figure R-12-1.

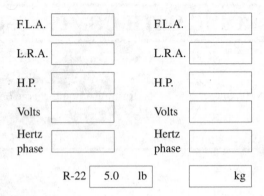

Figure R-12-1

STEP 2.

    A. Liquid charging is always much faster than charging with vapor.

    B. Liquid is charged on the high side of the system.

    C. On small systems, the charging is done with the compressor off. Since the full charge is seldom completed, the vapor method must be used to complete the process. You must also make sure that no liquid enters the compressor. Therefore most small systems are vapor charged.

    D. On large systems, a king valve located between the condenser and the metering valve offers a convenient means of charging the system on the high side, with the compressor running.

STEP 3.

    A. Purge the air from the lines while connecting the hoses from the gauge manifold to the high and low side of the refrigeration system (Refer to Laboratory Worksheet R-1 *Basic Refrigeration System Startup,* Step 4).

    B. Purge each line for about 2 s.

    C. It is always best to purge the lines with vapor and the cylinder in the upright position as shown in Figure R-12-3.

STEP 4. After the lines have been purged you are ready to begin charging as follows:

    A. Make sure that the high and low side gauge manifold valves are closed to start.

    B. Open the refrigerant cylinder fully and then turn it over and balance it on a digital scale.

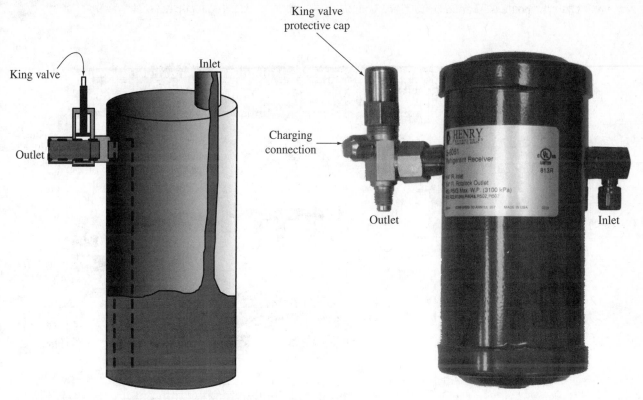

Figure R-12-2

King valve

Inlet

Outlet

King valve
protective cap

Charging
connection

Outlet

HENRY

Refrigerant Receiver

Inlet

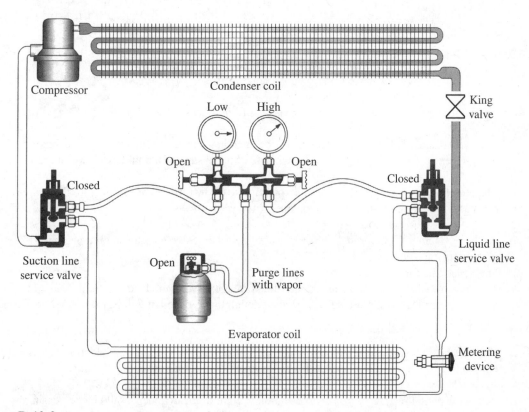

Compressor

Condenser coil

Low

High

King
valve

Open

Open

Closed

Closed

Suction line
service valve

Open

Purge lines
with vapor

Liquid line
service valve

Evaporator coil

Metering
device

Figure R-12-3

179

C. Once the bottle is inverted and balanced on the scale, set the digital readout on the scale to zero.

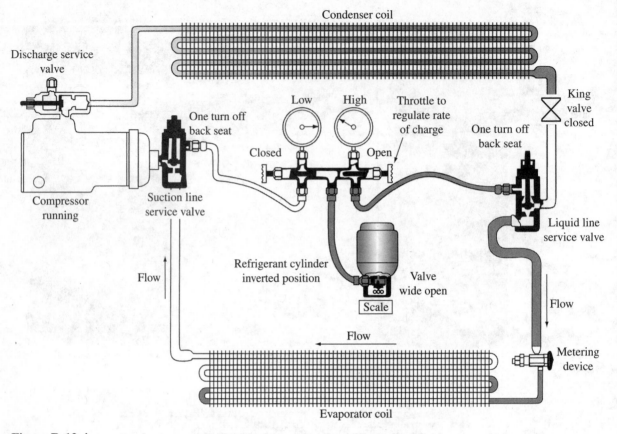

Figure R-12-4

D. Crack the suction and liquid line service valves off their back seats and *close the king valve.*

E. The system should already have been evacuated and in a vacuum. When the high side service valve is opened, the refrigerant will flow freely into the high side of the system as a liquid as shown in Figure R-12-5.

F. The metering device must be in the open position to allow refrigerant to flow into the evaporator. If the system has a box solenoid valve it must be open.

G. As the refrigerant flows through the evaporator, the suction side pressure should begin to rise above the low pressure cutout setting.

H. You can now start the system and run it in the normal operating mode and with the king valve closed, the liquid line service valve pressure will be low enough that refrigerant will be drawn from the cylinder into the system.

I. Throttle the high side gauge manifold valve to control the rate of charge and continue charging until the proper amount of refrigerant has been added as indicated by the reading on the digital scale.

J. Watch the digital readout as you charge and the weight of the cylinder should be decreasing by a negative amount. As an example, assume the system requires a 12 oz charge. You would continue to charge until the readout measured –12 oz (minus 12 oz).

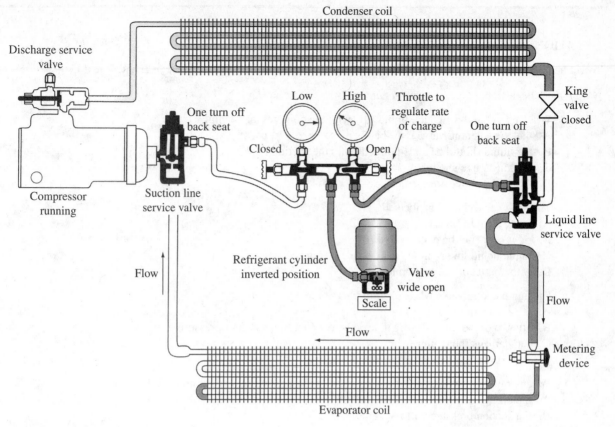

Figure R-12-5

STEP 5.

A. When the proper amount of refrigerant has been added to the system, close the gauge manifold high side valve and then *open the king valve.*

B. Even though the high side and low side gauge manifold valves are now both closed, you will still be able to monitor the refrigeration system pressures due to the configuration of the gauge manifold.

C. Keep the gauge manifold connected long enough to allow the system time to stabilize.

D. Verify that the suction and discharge pressures of the refrigeration system agree with the manufacturer's recommendations and record the pressures below.

   Suction pressure _____ Discharge pressure _____

E. After the system has been charged and the operating pressures verified, back seat the suction and discharge service valves and close the valve on the refrigerant cylinder.

F. Carefully disconnect the hoses.

QUESTIONS

To get help in answering some of the following questions, refer to the *Fundamentals of HVAC/R* text Unit 27.

(Circle the letter that indicates the correct answer.)

1. When charging through the liquid side with the compressor running:
   A. the box solenoid valve must be open.
   B. the liquid should be added directly into the suction.
   C. the king valve should be open.
   D. All of the above are correct.

2. The king valve should be located:
   A. in the suction line.
   B. right opposite the queen valve.
   C. in the liquid line right after the condenser or receiver.
   D. at the outlet of the evaporator.

3. Closing the king valve while charging:
   A. is bad practice.
   B. allows for the line pressure to drop below charging cylinder pressure.
   C. can damage the compressor.
   D. will trip the unit on the high pressure cutout.

4. When preparing to charge, if the king valve is closed but the refrigerant cylinder is not open:
   A. the discharge pressure will rise rapidly.
   B. the suction pressure will rise rapidly.
   C. Both A and B are correct.
   D. the unit will stop on the low pressure cutout.

LIQUID CHARGING WITH BLENDS (ZEOTROPES)

LABORATORY OBJECTIVE

The student will demonstrate the correct procedure for performing a liquid charge with a blended (zeotrope) refrigerant on a small refrigeration unit using a digital scale.

LABORATORY NOTES

Prior to charging, the system must be leak tested and evacuated (refer to Laboratory Worksheets R-5 and R-6 for the correct procedures for evacuation). In this way, when the charging is started, the system is under vacuum so that when the refrigerant enters the system it will be drawn into the unit due to the difference in pressure.

When charging refrigerants that fractionate (zeotropes), such as R-401a, they must be charged as a liquid to prevent separation of the refrigerant as it enters the system.

This can be done with a 100% liquid charge as described in Laboratory Worksheet R-12 with a refrigeration system that has a king valve. This can also be accomplished on smaller systems as described in the following procedure.

FUNDAMENTALS OF HVAC/R TEXT REFERENCE
Unit 27

Required Tools and Equipment

Gloves and goggles	Unit 3
Service valve wrench	Unit 9
Gauge manifold	Unit 12
Digital scale	Unit 27
Cylinder of refrigerant	
Operating refrigeration system	

SAFETY REQUIREMENTS

A. Wear safety goggles and gloves when working with refrigerants. Liquid refrigerant can cause frostbite when in contact with eyes and skin.

B. Use low loss hose fittings, or wrap cloth around hose fittings before removing the fittings from a pressurized system or cylinder. Inspect all fittings before attaching hoses.

PROCEDURE

STEP 1. The refrigerant type and required amount of charge can be found on the refrigeration system name plate as shown in Figure R-13-1.

F.L.A.			F.L.A.	
L.R.A.			L.R.A.	
H.P.			H.P.	
Volts			Volts	
Hertz phase			Hertz phase	

R-22   5.0   lb   kg

Figure R-13-1

STEP 2. Liquid charging is always the method used for charging blended refrigerants (zeotropes). Zeotropic blends generally consist of three different types of refrigerant blended together and they all have different boiling points. Blends can fractionate when allowed to vaporize which means that one of the refrigerants will boil off before the others. If you charge a blend as a vapor, the mixture of refrigerant entering the system will not be representative of the mixture in the blend.

STEP 3. Liquid is charged on the high side of the system. On small systems the charging is done with the compressor off and the full charge is seldom completed. Therefore additional charging is required to complete the process. You must also make sure that no liquid enters the compressor.

STEP 4.

    A. Place a refrigerant cylinder on a digital scale in the upright position and connect the gauge manifold as shown in Figure R-13-2.

    B. Purge the air from the lines while connecting the remaining hoses from the gauge manifold to the high and low side of the refrigeration system that you are charging (refer to Laboratory Worksheet R-1 *Basic Refrigeration System Startup,* Step 4).

    C. If the system has Schrader valves, then be careful not to allow air to enter the system when you connect the hoses.

STEP 5. After the lines have been purged you are ready to begin charging as follows:

    A. Normally the refrigerant cylinder must be turned upside down (inverted) for liquid.

    B. It is always good practice to open the cylinder valve all the way open prior to turning it over.

    C. Once the bottle is inverted and balanced on the scale, set the digital readout on the scale to zero.

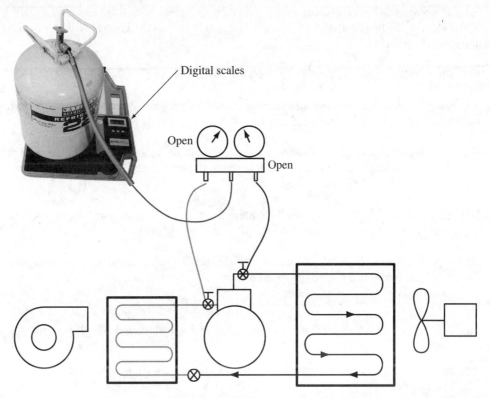

Figure R-13-2

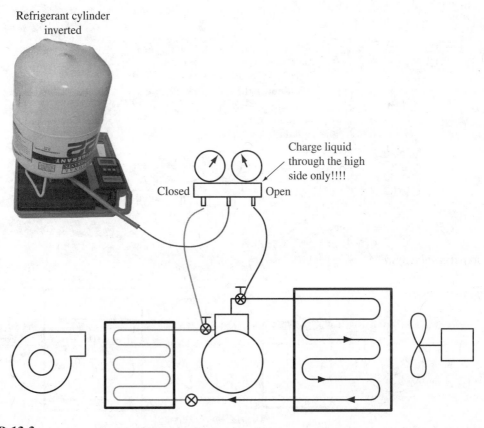

Figure R-13-3

D. Prior to opening the service valves on the refrigeration system, make sure that ONLY THE HIGH SIDE OF THE GAUGE MANIFOLD IS OPEN AND THE LOW SIDE GAUGE MANIFOLD VALVE IS CLOSED as shown in Figure R-13-3.

E. NEVER ADMIT LIQUID INTO THE SUCTION SIDE OF THE COMPRESSOR!!

F. The system should already have been evacuated and in a vacuum. When the high side service valve is opened, the refrigerant should flow freely into the high side of the system as a liquid.

G. Watch the digital readout as you charge and the weight of the cylinder should be decreasing by a negative amount. As an example, assume the system requires a 12oz charge. You would continue to charge until the readout measured –12oz (minus 12oz).

H. It is unlikely that you will overcharge the system at this point because normally only 50% to 75% of the charge will flow before the charging stops and the pressures equalize.

I. At this point close the high side valve on the gauge manifold and record the readout in pounds and ounces from the digital scale in the space provided below.

Amount of liquid charge _____lb_____oz_____

Step 6. If the full charge is not complete after adding the liquid, then additional charging must be performed to complete the process.

A. You must never allow liquid refrigerant to enter the compressor suction, however you must finish charging the blend as a liquid.

B. To accomplish this you will introduce liquid to the gauge manifold; however, you will throttle the valve to allow for only a very slight flow.

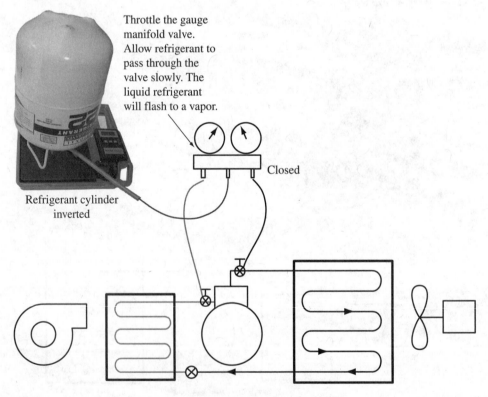

Throttle the gauge manifold valve. Allow refrigerant to pass through the valve slowly. The liquid refrigerant will flash to a vapor.

Closed

Refrigerant cylinder inverted

Figure R-13-4

C. The liquid refrigerant will flash to vapor as it passes through the gauge manifold valve.

D. Start the system in the normal run mode and the discharge pressure should register on the high side of the gauge manifold.

E. Although there is no flow through the high side of the gauge manifold, you will still be able to monitor the system discharge pressure on the high side gauge. See Figure R-13-4.

F. You will throttle the gauge manifold low side valve as you begin vapor charging. This will allow the liquid in the hose to flash so as not to introduce liquid into the suction side of the compressor.

G. Slowly open the refrigeration system suction service valve while throttling the low side gauge manifold valve.

H. When the proper amount of refrigerant has been added to the system, close the gauge manifold low side valve.

I. Even though the high side and low side gauge manifold valves are now both closed, you will still be able to monitor the refrigeration system pressures due to the configuration of the gauge manifold.

J. Keep the gauge manifold connected long enough to allow the system time to stabilize.

K. Verify that the suction and discharge pressures of the refrigeration system agree with the manufacturer's recommendations and record the pressures below.

Suction pressure _____ Discharge pressure _____

L. After the system has been charged and the operating pressures verified, back seat the suction and discharge service valves and close the valve on the refrigerant cylinder.

M. Carefully disconnect the hoses.

QUESTIONS

To get help in answering some of the following questions, refer to the *Fundamentals of HVAC/R* text Unit 27.

(Circle the letter that indicates the correct answer.)

1. Zeotropic blends:
   A. can experience "temperature glide."
   B. are numbered in the 500 series.
   C. have one single boiling point.
   D. All of the above are correct.

2. Fractionation occurs when:
   A. refrigerants are mixed with oil.
   B. charging blends as a liquid.
   C. charging blends as a vapor.
   D. None of the above is correct.

3. When liquid charging:
   A. never allow liquid refrigerant to enter the compressor.
   B. liquid refrigerant blends can be allowed to enter the compressor.
   C. always fractionate the refrigerant first.
   D. a weight measurement is not required.

4. Throttling liquid refrigerant through the gauge manifold when charging a liquid blend:
   A. will increase its temperature.
   B. will allow some of it to flash to vapor.
   C. will damage the gauge manifold.
   D. will damage the gauge manifold valve seats.

5. What types of refrigerants must be charged as a liquid?
   A. Azeotropic blends.
   B. Zeotropic blends.
   C. Near-azeotropic blends.
   D. Both B and C must be charged as liquids.

6. Zeotropic blends are identified as the:
   A. Z series.
   B. 500 series.
   C. 400 series.
   D. 700 series.

7. The weight of a closed cylinder containing a blend will vary with temperature:
   A. True.
   B. False.

SOLENOID VALVES

LABORATORY OBJECTIVE
The student will disassemble a solenoid valve and be able to describe how it operates.

LABORATORY NOTES
For this lab exercise there should be one or more solenoid valves available that may be disassembled.

Solenoid valves can be used for many applications in a refrigeration system. They can be used to regulate flow to the evaporator and as king valves. They can also be used for the defrost cycle as well as other refrigeration applications.

FUNDAMENTALS OF HVAC/R TEXT REFERENCE
Units 11, 29, and 78

Required Tools and Equipment

Solenoid valves	Unit 78
Multimeter	Unit 11

SAFETY REQUIREMENTS
None.

PROCEDURE

STEP 1. Locate a solenoid valve and examine it carefully so that you may complete the following exercise:

A. Record all of the data that can be found on the valve such as volts, cycles, wattage, manufacturer, etc., as demonstrated for the sample solenoid valve in Figure A-1-1.

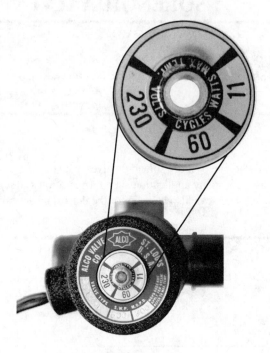

Figure A-1-1

Solenoid valve data: _____

_____

_____

_____

B. Disassemble the valve, being careful not to lose components while identifying the manner in which the valve came apart so that it may then be put back together again. See Figure A-1-2.

Coil

Plunger

Diaphragm

Valve body

Figure A-1-2

C. Sketch a cross sectional view of the disassembled valve in the space provided below.

D. Explain how the valve operates using the sketch you provided above.

_____

_____

_____

_____

_____

_____

_____

_____

STEP 2. Measure the coil resistance as follows:

A. The coil removed from the solenoid valve will have an identification number for spare parts purposes. Notice the part identification number on the coil shown in the illustration in Figure A-1-3.

Figure A-1-3

B. Using the coil part number and the manufacturer's data, you should be able to determine the normal resistance of the coil.

C. Familiarize yourself with a multimeter and zero it in (refer to Laboratory Worksheet E-1, *Electrical Meters*).

D. Connect one of the leads of the multimeter to each wire attached to the coil.

E. Start with the resistance reading on the highest scale.

F. Record the resistance reading _____Ω.

G. Compare this reading to the manufacturer's data to see if the coil is satisfactory.

STEP 3. Reassemble the solenoid valve, making sure that *none of the parts is lost and each is in its proper order.*

QUESTIONS

To get help in answering some of the following questions, refer to the *Fundamentals of HVAC/R* text Units 11, 29, and 78.

1. A common failure with solenoid valves is that the coil eventually burns out. Regarding the valve you disassembled, if the coil burned out, would the valve fail open or closed?

   _____

2. If the measured resistance of the coil is infinite, what would this indicate?

   _____

   _____

3. If the measured resistance of the coil is zero, what would this indicate?

   _____

   _____

(Circle the letter that indicates the correct answer.)

4. If the coil on the solenoid supplying refrigerant to an evaporator burned out:
   A. the refrigerated space would warm up.
   B. the refrigerated space would cool down.
   C. the compressor would shut down on high pressure.
   D. the compressor motor would overload.

5. If a solenoid valve is energized:
   A. it will always be closed.
   B. then the oil must be burned out.
   C. it may feel warm to the touch.
   D. then the circuit is faulty.

6. If the measured resistance of a solenoid coil is infinite:
   A. then the coil is satisfactory.
   B. the coil may have a short.
   C. the coil may have an open.
   D. the coil will run hot.

7. If the measured resistance of a solenoid coil is infinite:
   A. then the coil is satisfactory.
   B. the coil may have a short.
   C. the coil may have an open.
   D. the coil will run hot.

8. Many refrigerant solenoid valves often fail in the:
   A. neutral position.
   B. reverse flow position.
   C. open position.
   D. closed position.

9. When the coil of a solenoid valve is energized:
   A. it will spin.
   B. it will act as a magnet.
   C. its polarity will be reversed.
   D. it will slowly rotate.

10. An energized solenoid valve:
    A. will act as a magnet.
    B. may make a slight humming sound.
    C. may feel warm to the touch.
    D. All of the above are correct.

11. Many solenoid valves have a pilot orifice to assist in valve lift:
    A. to speed up the action of the valve.
    B. to reverse the flow through the valve.
    C. for manual operation if the coil burns out.
    D. to allow for smaller size coils to be used.

12. Before removing a coil from a solenoid valve:
    A. make sure that you have a spare.
    B. twist the ends of the wires together.
    C. tap it gently with a wrench.
    D. deenergize the circuit.

INTERNALLY EQUALIZED THERMOSTATIC EXPANSION VALVES

LABORATORY OBJECTIVE

The student will disassemble an internally equalized thermostatic expansion valve and be able to describe how it operates.

LABORATORY NOTES

For this lab exercise there should be one or more internally equalized thermostatic expansion valves available that may be disassembled.

FUNDAMENTALS OF HVAC/R TEXT REFERENCE

Unit 21

Required Tools and Equipment

Internally equalized thermostatic expansion valve	Unit 21

SAFETY REQUIREMENTS

None.

PROCEDURE

STEP 1. Locate an internally equalized thermostatic expansion valve and examine it carefully so that you may complete the following exercise (refer to the *Fundamentals of HVAC/R* text Unit 21 for additional information):

    A. Record all of the data that can be found on the valve such as refrigerant type, capacity, line size, refrigerant bulb charge, etc., such as the valves shown in Figure A-2-1 and Figure A-2-2.

Thermal bulb

Figure A-2-1

Thermostatic expansion valve data: _____

_____

_____

B. Disassemble the valve being careful not to lose components while identifying the manner in which the valve came apart so that it may then be put back together again. An internally equalized valve has only an inlet and an outlet. There is no equalizing connection.

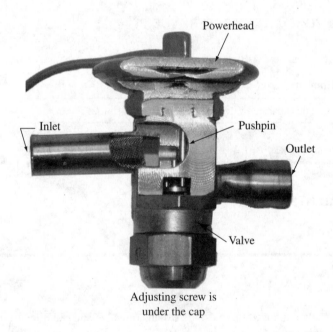

Powerhead

Inlet

Pushpin

Outlet

Valve

Adjusting screw is
under the cap

Figure A-2-2

C. Sketch a cross sectional view of the disassembled valve in the space provided below.

D. Explain how the valve operates using the sketch you provided above.

_____

_____

_____

_____

_____

_____

_____

_____

STEP 2. Using the disassembled valve, your sketch, and your description, answer the following questions:

A. Draw a one-line representation of the diaphragm and the forces acting upon it (bulb pressure, evaporator pressure, and spring pressure).

B. How is the diaphragm movement transmitted to the valve disk?

_____

_____

C. Does the spring help to open or to close the valve?

_____

D. How does the evaporator pressure travel to the underside of the valve diaphragm?

_____

_____

E. Does the evaporator pressure help to open or to close the valve?

_____

F. If the bulb lost its charge would the valve fail open or closed?

_____

G. What type of refrigerant can the valve be used for?

_____

QUESTIONS

To get help in answering some of the following questions, refer to the *Fundamentals of HVAC/R* text Unit 21.

(Circle the letter that indicates the correct answer.)

1. A thermostatic expansion valve can be adjusted for:
   A. refrigerant flow through the evaporator.
   B. evaporator refrigerant outlet superheat.
   C. evaporator capacity.
   D. All of the above are correct.

2. An internally equalized thermostatic expansion valve senses pressure:
   A. at the evaporator outlet.
   B. at the compressor inlet.
   C. at the condenser outlet.
   D. at the evaporator inlet.

3. To increase the superheat setting of a thermostatic expansion valve you would turn the adjusting screw:
   A. clockwise.
   B. counterclockwise.
   C. It cannot be adjusted.
   D. back and forth.

4. If the thermal bulb on a thermostatic expansion valve lost its charge:
   A. the valve would open wide.
   B. the valve would hunt.
   C. the valve would close.
   D. the valve would frost up.

5. The thermal bulb pressure on the top of a thermostatic expansion valve diaphragm is transmitted to the valve by:
   A. the spring.
   B. the evaporator inlet pressure.
   C. the evaporator outlet pressure.
   D. pushpins.

6. The thermal bulb force acting on the topside of the diaphragm in a thermostatic expansion valve is balanced by:
   A. spring force only.
   B. evaporator pressure only.
   C. spring force and evaporator pressure.
   D. an orifice plate.

7. Internally equalized thermostatic expansion valves can be used:
   A. on evaporator coils that have a minimal pressure drop.
   B. on evaporator coils that have a large pressure drop.
   C. on beverage coolers only.
   D. Both A and C are correct.

8. If an internally equalized thermostatic expansion valve is used on an evaporator coil that has a large pressure drop:
   A. the evaporator coil will flood with refrigerant.
   B. the evaporator coil will freeze up.
   C. the evaporator coil will be starved.
   D. liquid will slug back to the compressor.

EXTERNALLY EQUALIZED THERMOSTATIC EXPANSION VALVES

LABORATORY OBJECTIVE

The student will disassemble an externally equalized thermostatic expansion valve and be able to describe how it operates.

LABORATORY NOTES

For this lab exercise there should be one or more externally equalized thermostatic expansion valves available that may be disassembled.

FUNDAMENTALS OF HVAC/R TEXT REFERENCE
Unit 21

Required Tools and Equipment

Externally equalized thermostatic expansion valve	Unit 21

SAFETY REQUIREMENTS
None.

PROCEDURE

STEP 1. Locate an externally equalized thermostatic expansion valve and examine it carefully so that you may complete the following exercise (refer to the *Fundamentals of HVAC/R* text Unit 21 for additional information):

    A. Record all of the data that can be found on the valve such as refrigerant type, capacity, line size, refrigerant bulb charge, etc. See Figure A-3-1.

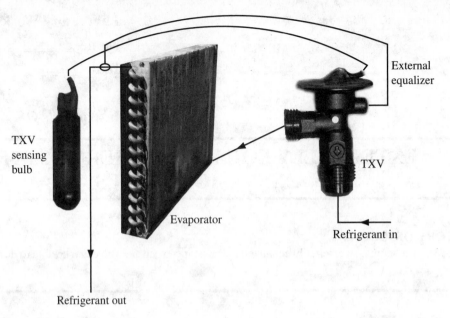

External
equalizer

TXV
sensing
bulb

TXV

Refrigerant in

Evaporator

Refrigerant out

Figure A-3-1

Thermostatic expansion valve data: _____

_____

_____

B. Disassemble the valve being careful not to lose components while identifying the manner in which the valve came apart so that it may then be put back together again. An externally equalized valve has an inlet, an outlet, and an external equalizing connection. See Figure A-3-2.

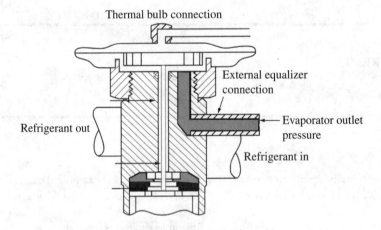

Thermal bulb connection

External equalizer
connection

Refrigerant out

Evaporator outlet
pressure

Refrigerant in

Figure A-3-2

C. Sketch a cross sectional view of the valve in the space provided below.

D. Explain how the valve operates using the sketch you provided above.

_____

_____

_____

_____

_____

_____

_____

_____

STEP 2. Using the disassembled valve, your sketch, and your description, answer the following questions:

A. Draw a one-line representation of the diaphragm and the forces acting upon it (bulb pressure, evaporator pressure, and spring pressure).

B. How is the diaphragm movement transmitted to the valve disk?

_____

_____

C. Does the spring help to open or to close the valve?

_____

D. How does the evaporator pressure travel to the underside of the valve diaphragm?

_____

_____

E. Does the evaporator pressure help to open or to close the valve?

_____

F. If the bulb lost its charge would the valve fail open or closed?

_____

G. What type of refrigerant can the valve be used for?

_____

QUESTIONS

To get help in answering some of the following questions, refer to the *Fundamentals of HVAC/R* text Unit 21.

(Circle the letter that indicates the correct answer.)

1. A thermostatic expansion valve can be adjusted for:
   A. refrigerant flow through the evaporator.
   B. evaporator refrigerant inlet superheat.
   C. evaporator pressure.
   D. All of the above are correct.

2. An externally equalized thermostatic expansion valve senses pressure:
   A. at the evaporator outlet.
   B. at the compressor inlet.
   C. at the condenser outlet.
   D. at the evaporator inlet.

3. To decrease the superheat setting of a thermostatic expansion valve you would turn the adjusting screw:
   A. clockwise.
   B. counterclockwise.
   C. It cannot be adjusted.
   D. back and forth.

4. If the thermal bulb on a thermostatic expansion has a liquid charge:
   A. control will always be from the bulb.
   B. control will never be from the bulb.
   C. the valve will have a limited opening.
   D. the valve will open quicker.

5. If the thermal bulb on a thermostatic expansion has a limited liquid (gas) charge:
   A. control will always be from the bulb.
   B. control will never be from the bulb.
   C. the valve will have a limited opening.
   D. the valve will open quicker.

6. If the thermal bulb on a thermostatic expansion has a cross charge:
   A. the fluid in the bulb is the same as the refrigerant in the system.
   B. the fluid in the bulb is different than the refrigerant in the system.
   C. the suction pressure crosses the discharge pressure.
   D. the valve will open quicker.

7. Externally equalized thermostatic expansion valves can be used:
   A. on evaporator coils that have a minimal pressure drop.
   B. on evaporator coils that have a large pressure drop.
   C. on beverage coolers only.
   D. Both A and C are correct.

8. If the external equalizing connection on an externally equalized thermostatic expansion valve is capped closed:
   A. the evaporator coil will flood with refrigerant.
   B. the evaporator coil will freeze up.
   C. the evaporator coil will be starved.
   D. liquid will slug back to the compressor.

9. The fluid in a straight charged thermostatic expansion valve bulb:
   A. is the same as the refrigerant in the system.
   B. is different from the refrigerant in the system.
   C. is always liquid.
   D. is always vapor.

ADJUSTING THE WATER REGULATING VALVE

LABORATORY OBJECTIVE
The student will demonstrate how to properly adjust the set point on a water regulating valve for a refrigeration system that has a water cooled condenser.

LABORATORY NOTES
For this lab exercise there needs to be an operating refrigeration system that has a water cooled condenser. Additionally the discharge pressure of the unit must be controlled by a water regulating valve.

FUNDAMENTALS OF HVAC/R TEXT REFERENCE
Units 20, 23, and 78

Required Tools and Equipment

Operating refrigeration unit with water cooled condenser and regulating valve	Unit 20
Installed or external thermometers	Unit 12
Gauge manifold	Unit 12

SAFETY REQUIREMENTS

    A. Always read the equipment manual to become familiarized with the refrigeration system and its accessory components prior to startup.

    B. Wear safety goggles and gloves when working with refrigerants. Liquid refrigerant can cause frostbite when in contact with eyes and skin.

    C. Use low loss hose fittings, or wrap cloth around hose fittings before removing the fittings from a pressurized system or cylinder. Inspect all fittings before attaching hoses.

PROCEDURE

STEP 1. Familiarize yourself with the major components in the refrigeration system including the condenser, compressor, evaporator, and metering device. Determine where the high side and low side connections for the system are located. In earlier applications of water cooled condensers, it was common practice to tap the water supply and then waste the discharge water to a drain connection. The cost and scarcity of water (unless drawn from a lake or wells and then retuned) has become prohibitive and it is even outlawed for refrigeration and air conditioning use by many local codes. Many installations today have water supplied to the condenser from a cooling tower and returned to the tower to dissipate the heat. It is important to identify the source of the cooling water because the temperature of this incoming water affects the condenser performance. In the system depicted in Figure C-1-1, the cooling water tower rejects heat picked up in the condenser to the outside air. Water cooled condensers permit about a 15°F lower compressor discharge pressure than air cooled units.

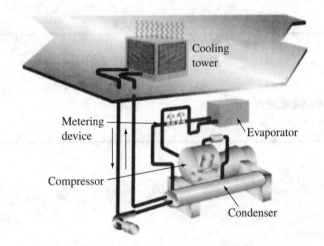

Figure C-1-1

STEP 2. Refrigerants all have different operating pressures and temperatures. An example is provided below of a system that has a refrigerant condensing temperature of 105°F. The water entering the condenser would be about 20°F cooler than that and would increase by about 10°F as it passes through the condenser.

For example, using R-22 and a 105°F condensing temperature, you should be able to determine from the P-T chart, Table R-1-1, depicted in Laboratory Worksheet R-1 *Basic Refrigeration System Startup* and from Figure 23-2 in Unit 23 of the *Fundamentals of HVAC/R* text that the condenser pressure (high side gauge) should read approximately 211 psig. For another type of refrigerant such as R-134a, the expected condenser pressure would be lower, approximately 135 psig.

A. Calculate the expected condenser pressure based upon the refrigerant type for the system you are working on.

B. If there are not thermometers currently on the unit then you must instrument the refrigeration system with thermometers to read the cooling water inlet, cooling water outlet, and condenser temperature as shown in Figure C-1-2.

C. If you are unable to place probes directly into the fluid stream flow, you can use pipe surface temperatures, however you must adjust for the heat transfer difference through the pipe.

D. If there are no pressure gauges currently on the unit then you must connect a gauge manifold to the high and low side of the system as described in Laboratory Worksheet R-1 *Basic Refrigeration System Startup*, Steps 4, 5, and 6.

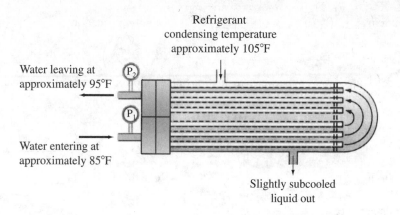

Refrigerant
condensing temperature
approximately 105°F

Water leaving at
approximately 95°F

Water entering at
approximately 85°F

Slightly subcooled
liquid out

Figure C-1-2

STEP 3. Adjust the water regulating valve as follows:

A. The water regulating valve operates with a spring force opposing the refrigerant pressure. As the refrigerant pressure increases, the valve will open more against the spring force and allow more water to flow through the valve. The forces will be balanced when the pressure returns to its setpoint.

B. An increased spring force will lead to an increased condenser pressure. Many water regulating valves are marked with an indicator to identify in which direction to turn the adjusting screw. See Figure C-1-3.

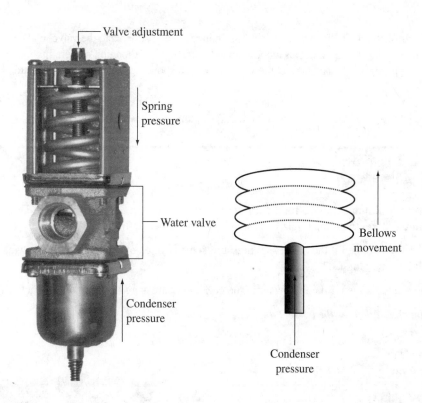

Valve adjustment

Spring
pressure

Water valve

Condenser
pressure

Bellows
movement

Condenser
pressure

Figure C-1-3

STEP 4. Start the refrigeration system by first turning on the cooling water flow and record the following:

Running

High side pressure _____   Low side pressure _____

Cooling water in _____°F   Cooling water out _____°F

Condenser temperature _____°F

STEP 5. Turn the adjusting screw on the water regulating valve to lower the condenser pressure by 5 psig and record the following:

Running

High side pressure _____   Low side pressure _____

Cooling water in _____°F   Cooling water out _____°F

Condenser temperature _____°F

STEP 6. Turn the adjusting screw on the water regulating valve to increase the condenser pressure by 5 psig and record the following:

Running

High side pressure _____   Low side pressure _____

Cooling water in _____°F   Cooling water out _____°F

Condenser temperature _____°F

STEP 7.

A. When finished with taking the data from the changes in adjustments to the valve, make one final adjustment to set the valve for normal operation as recommended in the refrigeration unit's operating manual. If the unit does not have an operating manual, then set the valve for a corresponding condenser temperature of approximately 105°F.

B. Allow the system to run and stabilize prior to shutting down and then carefully disconnect the gauge manifold and any instrumentation that you attached to the unit.

QUESTIONS

1. Compare your temperatures for the different valve settings in Steps 4 through 7. Does the cooling water temperature differential remain fairly constant over the varied settings?

2. What would happen if the condenser pressure sensing line connected to the valve was accidentally pinched so that it became plugged?

3. What would happen if the bellows that the condenser pressure is acting upon cracked?

(Circle the letter that indicates the correct answer.)

4. The water regulating valve would fail:
   A. wide open.
   B. closed.
   C. only if the electrical contact was open.
   D. if the water supply was increased.

5. If the water regulating valve failed:
   A. the compressor would short cycle.
   B. the compressor would start on the low pressure cutout.
   C. the compressor would stop on the high pressure cutout.
   D. nothing would happen to the system.

SETTING THE HIGH PRESSURE CUTOUT

LABORATORY OBJECTIVE

The student will demonstrate how to properly set the cutout and cut-in points on a high pressure cutout switch.

LABORATORY NOTES

For this lab exercise there needs to be an operating refrigeration system that has a high pressure cutout switch. It can be set to stop the compressor before excessive pressures are reached. Such conditions might occur because of a water supply failure in water cooled condensers or because of a fan motor stoppage on air cooled condensers.

FUNDAMENTALS OF HVAC/R TEXT REFERENCE
Unit 78

Required Tools and Equipment

Operating refrigeration unit with high pressure cutout switch	Unit 78
Gauge manifold	Unit 12

SAFETY REQUIREMENTS

    A. Always read the equipment manual to become familiarized with the refrigeration system and its accessory components prior to startup.

    B. Wear safety goggles and gloves when working with refrigerants. Liquid refrigerant can cause frostbite when in contact with eyes and skin.

    C. Use low loss hose fittings, or wrap cloth around hose fittings before removing the fittings from a pressurized system or cylinder. Inspect all fittings before attaching hoses.

PROCEDURE

STEP 1. Familiarize yourself with the major components in the refrigeration system including the condenser, compressor, evaporator, and metering device. Determine where the high side and low side connections for the system are located.

A. The operating manual with the refrigeration specifications should indicate what the expected discharge pressure should be. If not, then you must calculate the expected discharge pressure based upon the refrigerant type for the system you are working on.

B. For air cooled condensers, the refrigerant condensing temperature is typically 30°F higher than ambient (room) temperature. If it is 90°F outside, then the refrigerant condensing temperature will equal 90°F + 30°F = 120°F.

   You should be able to determine the condenser pressure (high side gauge) from the P-T chart, Table R-1-1, depicted in Laboratory Worksheet R-1 *Basic Refrigeration System Startup* and from Figure 23-2 in Unit 23 of the *Fundamentals of HVAC/R* text.

   For R-22 the pressure would be expected to be approximately 260 psig. For another type of refrigerant such as R-134a, the expected condenser pressure would be somewhat lower, approximately 172 psig.

C. For water cooled condensers, the refrigerant condensing temperature is typically 10°F higher than the water leaving the condenser. If the water temperature is 85°F entering the condenser and 95°F leaving the condenser, then the refrigerant condensing temperature will equal 95°F + 10°F = 105°F.

   You should be able to determine the condenser pressure (high side gauge) from the P-T chart, Table R-1-1, depicted in Laboratory Worksheet R-1 *Basic Refrigeration System Startup* and from Figure 23-2 in Unit 23 of the *Fundamentals of HVAC/R* text.

   For R-22 the pressure would be expected to be approximately 211 psig. For another type of refrigerant such as R-134a, the expected condenser pressure would be somewhat lower, approximately 135 psig.

STEP 2. Determine the high pressure cutout setting as follows (see Figure C-2-1 and Figure C-2-2):

A. The high pressure cutout is typically set at 125% of normal discharge pressure.

B. For an air cooled condenser using R-22, the setting would be approximately 1.25 × 260 psig = 325 psig. For R-134a it would be 1.25 × 172 psig = 215 psig.

C. For a water cooled condenser using R-22, the setting would be approximately 1.25 × 211 psig = 264 psig. For R-134a it would be 1.25 × 135 psig = 169 psig.

D. Depending on the type of control switch, the high pressure cutout may reset automatically once the pressure reaches the cut-in value (this would be at some point below the normal discharge pressure—in this example about 100 psig).

E. Remember, cycling on the high pressure cutout is *harmful* to the unit!

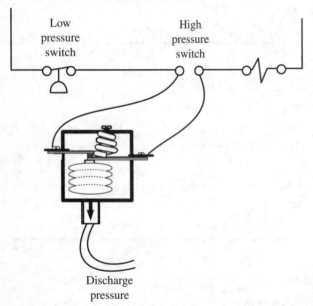

Figure C-2-1

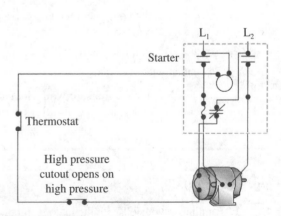

Figure C-2-2

STEP 3.

    A. Adjust the high pressure cutout setting by turning the adjusting screw to the correct setting as determined from Step 2.

    B. Also adjust the cut-in setting if the high pressure cutout is so equipped. See Figure C-2-3.

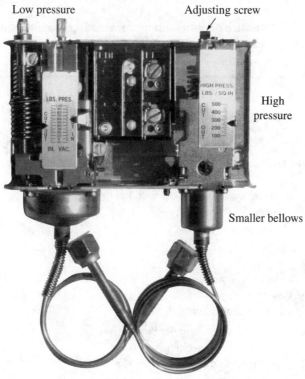

Dual high pressure and low pressure cutout

Figure C-2-3

STEP 4.

    A. If there are no pressure gauges currently on the unit then you must connect a gauge manifold to the high and low side of the system as described in Laboratory Worksheet R-1 *Basic Refrigeration System Startup*, Steps 4, 5, and 6.

    B. Start the refrigeration system in the normal cooling mode and record the following:

    Running

    High side pressure _____

    Low side pressure _____

STEP 5.

    A. After the system has stabilized, turn off the cooling air or water and allow the system to cut out on high pressure. Record the cutout pressure below.

    High side cutout pressure _____

B. Check the cut-in pressure if the switch is so equipped. Turn the cooling air or water back on and the discharge pressure should begin to drop. The unit should restart at the high pressure cutout-cut-in setting. Record the cut-in pressure below.

High side cut-in pressure _____

STEP 6.

A. Continue making any necessary adjustments until the high pressure cutout is correctly set.

B. Allow the system to run and stabilize prior to shutting down and then carefully disconnect the gauge manifold and any instrumentation that you attached to the unit.

QUESTIONS

(Circle the letter that indicates the correct answer.)

1. The high pressure cutout switch:
   A. is a safety device.
   B. is a device to cycle the compressor on and off with load.
   C. contacts are normally open.
   D. contacts close on high pressure.

2. The cut-in setting on a high pressure cutout switch.
   A. bypasses the cutout portion.
   B. restarts the compressor when the pressure is back to normal.
   C. shuts off the compressor on high pressure.
   D. restarts the compressor on high pressure.

3. The high pressure cutout should be set to stop the compressor at:
   A. a pressure that is 75% above normal discharge pressure.
   B. a pressure that is 25% above normal discharge pressure.
   C. a pressure that is 125% above normal discharge pressure.
   D. a pressure that is 150% above normal discharge pressure.

4. Cycling of the compressor on the high pressure cutout:
   A. is always desirable.
   B. is unavoidable.
   C. is normal.
   D. is harmful to the compressor.

5. A refrigeration system with a normal discharge pressure of 120 psig:
   A. would have a high pressure cutout setting of 125 psig.
   B. would have a high pressure cutout setting of 150 psig.
   C. would have a high pressure cutout setting of 135 psig.
   D. would have a high pressure cutout setting of 185 psig.

6. The cutout setting on a high pressure cutout switch:
   A. bypasses the cutout portion.
   B. restarts the compressor when the pressure is back to normal.
   C. shuts off the compressor on high pressure.
   D. restarts the compressor on high pressure.

7. Two operating conditions that could lead to the compressor stopping on the high pressure cutout are:
   A. lack of cooling air or cooling water.
   B. high current.
   C. a shorted winding.
   D. All of the above are correct.

SETTING THE LOW PRESSURE CUTOUT

LABORATORY OBJECTIVE
The student will demonstrate how to properly set the cutout and cut-in points on a low pressure cutout switch.

LABORATORY NOTES
For this lab exercise there needs to be an operating refrigeration system that has a low pressure cutout switch.

The low pressure switch, which senses compressor suction pressure, opens on a drop in pressure. It is set to cut out at a protective low limit pressure, but remains closed at normal operating pressures. It is also known as a loss of charge or low pressure cutout switch.

FUNDAMENTALS OF HVAC/R TEXT REFERENCE
Unit 78

Required Tools and Equipment

Operating refrigeration unit with low pressure cutout switch and king valve	Unit 78
Gauge manifold	Unit 12

SAFETY REQUIREMENTS
A. Always read the equipment manual to become familiarized with the refrigeration system and its accessory components prior to startup.

B. Wear safety goggles and gloves when working with refrigerants. Liquid refrigerant can cause frostbite when in contact with eyes and skin.

C. Use low loss hose fittings, or wrap cloth around hose fittings before removing the fittings from a pressurized system or cylinder. Inspect all fittings before attaching hoses.

PROCEDURE

STEP 1. Familiarize yourself with the major components in the refrigeration system including the condenser, compressor, evaporator, and metering device. Determine where the high side and low side connections for the system are located. See Figure C-3-1 and Figure C-3-2.

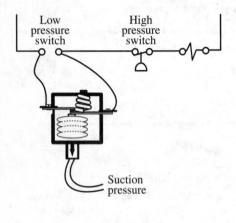

Figure C-3-1

Figure C-3-2

A. The operating manual with the refrigeration specifications should indicate what the expected suction pressure should be. Many low pressure control settings exist and different situations call for different settings.

B. If you have no information on the system, you may be able to calculate the expected suction pressure based upon the refrigerant type for the system you are working on.

C. The refrigerant temperature in the evaporator should generally be no more than 15°F colder than the medium being cooled.

D. As an example, if a freeze box is kept at 0°F, then the refrigerant temperature should be about –15°F.

   You should be able to determine the lowest evaporator pressure from the P-T chart, Table R-1-1, depicted in Laboratory Worksheet R-1 *Basic Refrigeration System Startup* and from Figure 23-2 in Unit 23 of the *Fundamentals of HVAC/R* text.

   For R-22 it would be expected to be approximately 13 psig. If the space to be cooled was for vegetables to be kept at 40°F, then the corresponding refrigerant temperature would be 40°F − 15°F = 25°F, which corresponds to an evaporator pressure of 49 psig.

STEP 2. Determine the low pressure cutout setting as follows:

A. If the low pressure cutout is in place to control the space temperature, then first determine the maximum and minimum temperature of the space. As an example, a space is to be kept at a temperature of between 40°F to 50°F. As explained in the previous section, the refrigerant temperature would be approximately 15°F colder than the space and would therefore vary between 25°F to 35°F. Using R-22, this would correspond to an evaporator pressure of from 49 to 62 psig.

The low pressure cutout would be set to start the compressor at the warmer temperature and higher pressure of 62 psig and then stop the compressor at the lower temperature and lower pressure of 49 psig.

In this manner, the low pressure cutout would cycle the compressor on and off dependent on the cooling load.

B. If the low pressure cutout is in place for low charge protection, then first determine the minimum pressure allowed for the system. In this type of system, the space temperature is generally maintained by another control such as a box solenoid valve. As an example, many refrigeration systems operate at positive pressures to reduce the possibility of air being drawn in. In this case, the low pressure cutout is often set at some point slightly above atmospheric pressure.

A common setting for this type of low pressure cutout would be to stop the compressor at a low load and low pressure of 2 psig and then restart the compressor as the space warms up and the pressure rises to about 8 psig.

In this manner, the refrigerant pressure in the system is never allowed to fall below atmospheric pressure.

C. If the system loses its refrigerant charge, the space warms up and the compressor continuously runs. Due to a lack of refrigerant circulating through the compressor and its continuous running, the compressor and motor may be damaged.

In this type of undercharged situation, the suction pressure would continue to drop. If the system is equipped with a low pressure cutout, the system would automatically shut down at the system cutout pressure.

D. Depending on the type of refrigeration system, a low pressure cutout can operate as both a temperature control and a low charge protection control.

STEP 3. Adjust the low pressure cutout setting as determined from Step 2 by turning the adjusting screws as follows (see Figure C-3-3):

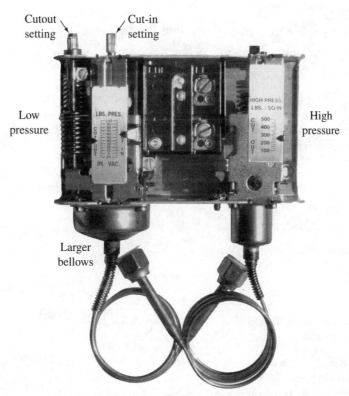

Dual high pressure and low pressure cutout

Figure C-3-3

A. Set the range adjustment. This is the difference between the minimum and maximum operating pressures within which the control will function. As an example, a cutout of 2 psig and a cut-in of 8 psig would indicate a typical range setting of a low pressure cutout.

B. Set the differential adjustment. This is the difference between the cutout and cut-in pressures for the control. The differential of a low pressure switch set to cut out at 2 psig and cut in at 8 psig would be 6 psig.

C. A low pressure cutout will have two adjusting screws as shown in Figure C-3-3. In many cases, one adjusting screw will control the range while the other controls the differential.

   Be careful as you make adjustments to the control. In some cases, the set range may change as you change the differential and vice versa.

STEP 4.

A. If there are no pressure gauges currently on the unit then you must connect a gauge manifold to the high and low side of the system as described in Laboratory Worksheet R-1 *Basic Refrigeration System Startup*, Steps 4, 5, and 6.

B. Start the refrigeration system in the normal cooling mode and allow the system to stabilize.

C. Close the king valve, which is located in the liquid line directly following the receiver or condenser. The suction pressure will begin to decrease and eventually the compressor will stop. Record the suction pressure at which the compressor stopped below:

   Suction pressure at which compressor stops _____

D. After recording the suction pressure at which the compressor stops, slowly open the king valve and observe the suction pressure begin to rise. The compressor should restart. Record the suction pressure at which the compressor restarts below:

   Suction pressure at which compressor restarts _____

STEP 5.

A. Continue making any necessary adjustments until the low pressure cutout is correctly set.

B. Allow the system to run and stabilize prior to shutting down and then carefully disconnect the gauge manifold and any instrumentation that you attached to the unit.

QUESTIONS

(Circle the letter that indicates the correct answer.)

1. The low pressure cutout switch:
   A. is a safety device.
   B. is a device to cycle the compressor on and off with load.
   C. contacts are normally open.
   D. Both A and B are correct.

2. The cut-in setting on a low pressure cutout switch:
   A. bypasses the cutout portion.
   B. restarts the compressor when the pressure is low.
   C. shuts off the compressor on a rise in suction pressure.
   D. restarts the compressor on a rise in suction pressure.

3. The low pressure cutout should be set to stop the compressor:
   A. when the desired box temperature is reached.
   B. before the suction pressure drops below atmospheric.
   C. Both A and B are correct.
   D. None of the above is correct.

4. The differential can be defined as the difference between the cutout and cut-in points of the control:
   A. True.
   B. False.

SETTING A BOX THERMOSTAT CONTROL

LABORATORY OBJECTIVE
The student will demonstrate how to properly set a thermostat that is controlling the temperature of a refrigerated space.

LABORATORY NOTES
For this lab exercise there needs to be an operating refrigeration system that has a thermostat controlling the temperature of the space being cooled.

FUNDAMENTALS OF HVAC/R TEXT REFERENCE
Unit 78

Required Tools and Equipment

Operating refrigeration unit with thermostat control	Unit 78

SAFETY REQUIREMENTS
Always read the equipment manual to become familiarized with the refrigeration system and its accessory components prior to startup.

PROCEDURE

STEP 1. Familiarize yourself with the major components in the refrigeration system including the condenser, compressor, evaporator, and metering device.

    A. *Temperature motor control*   This is the simplest type of control system. For this type of system there is only one space being cooled. The thermostat will cycle the compressor motor on and off dependent on the load. This type of system is shown in Figure C-4-1.

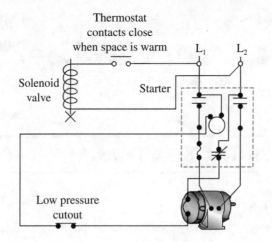

Figure C-4-1

B. *Pressure motor control*   This is used on systems that cool multiple spaces at one time. Each individual space will have its own solenoid valve, which is controlled by a thermostat. The low pressure cutout will cycle the compressor motor on and off dependent on the load. See Figure C-4-2.

C. The thermostat will have a sensing bulb that should be located in the space being cooled. The sensing bulb should be in such a location that it is sensing the average temperature of the space. If it is in a walk-in cooler, it should be mounted in a central location. Care should be taken not to damage the thin interconnecting tube located between the sensing bulb and the thermostat contacts. See Figure C-4-3.

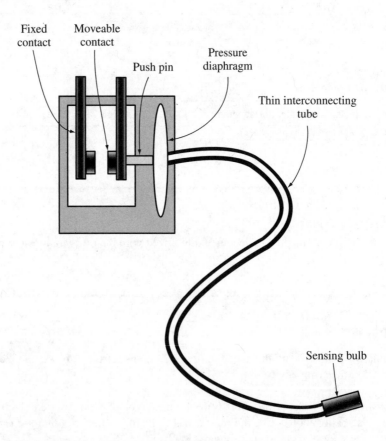

Figure C-4-2

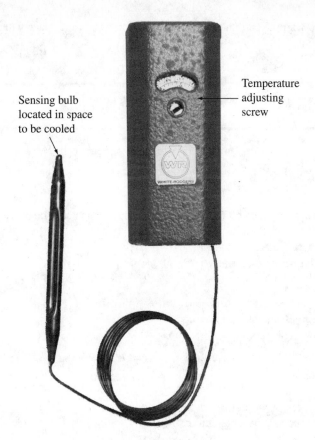

Sensing bulb
located in space
to be cooled

Temperature
adjusting
screw

Figure C-4-3

STEP 2. Set the thermostat as follows:

A. Determine the desired temperature for the space to be cooled and check the thermostat setting.

B. Generally the thermostat can be adjusted with a screwdriver or a small wrench, as seen in Figure C-4-3. Some thermostats have a readable scale for temperature. Others are adjusted by counting the number of turns of the screw. As an example, one full turn of the adjusting screw may equal 4°F, and so on.

C. *Differential adjustment* This is the difference between the cutout and cut-in pressures for the control. On a temperature type motor control, this setting will reduce the cycling of the compressor motor on and off. On a pressure type motor control, this setting will reduce the cycling of the solenoid valve open and closed.

As an example, assume the space is to be kept at a temperature of 33°F with a 4°F differential. With a temperature motor control, the compressor will start when the space temperature reaches 35°F and stop again some time later when the space cools back down to 31°F. This provides for an average temperature of 33°F.

D. Some spaces are allowed higher differentials than others. For example, dairy cooler temperatures should be fairly constant, without large variations in temperature. Therefore they would be set for a smaller differential.

STEP 3.

A. After setting the thermostat, start the refrigeration system in the normal cooling mode and allow the system to stabilize.

B. Depending on the size of the unit, it may take some time before the setpoint temperature is reached. Record the temperature of the space when the compressor stops or the solenoid valve closes.

_____°F

C. Allow the space to warm up and then record the temperature of the space when the compressor starts or the solenoid valve opens.

_____°F

D. Continue making any necessary adjustments until the thermostat is correctly set.

QUESTIONS

(Circle the letter that indicates the correct answer.)

1. If the thin interconnecting tube between the thermostat contacts and the sensing bulb is damaged:
   A. the compressor will not start.
   B. the space temperature will be too low.
   C. the system will operate as normal.
   D. the compressor will not stop.

2. A high differential setting on a thermostat:
   A. will lead to decreased cycling.
   B. will lead to lower than normal temperatures.
   C. will lead to increased cycling.
   D. will lead to higher than normal temperatures.

3. A temperature type motor control:
   A. cycles the compressor motor on and off dependent on load.
   B. is connected to the low pressure cutout.
   C. protects the motor from overheating.
   D. unloads the compressor at high space temperatures.

4. Generally the thermostat sensing bulb is located:
   A. right at the entrance to the space.
   B. connected to the evaporator coil.
   C. at a central location to sense average temperature.
   D. at the lowest point in the space.

5. The thermostats for a multiple space system that uses a pressure type motor control:
   A. cycle the compressor motor on and off dependent on load.
   B. are connected to the low pressure cutout.
   C. protect the motor from overheating.
   D. control individual space solenoid valves.

SETTING AN EVAPORATOR PRESSURE REGULATOR

LABORATORY OBJECTIVE

The student will demonstrate how to properly set an evaporator pressure regulator that is controlling the temperature of an evaporator coil for a refrigerated space.

LABORATORY NOTES

For this lab exercise there needs to be an operating refrigeration system that has an evaporator pressure regulator on the outlet of an evaporator for the space being cooled. If there are no thermometers or pressure gauges currently on the unit then you must instrument the refrigeration system using a gauge manifold to measure the evaporator pressure and a thermometer to read the evaporator refrigerant temperature.

If you are unable to place probes directly into the refrigerant stream flow, you can use evaporator coil surface temperatures; however, you must adjust for the heat transfer difference through the coil.

Multiple evaporator systems are commonly found in supermarkets. A single compressor may be used to control a number of different case or fixture temperatures.

Without an evaporator pressure regulator (EPR) all of these spaces would have a common evaporator pressure and temperature. The refrigerant would need to be cold enough for the low temperature boxes but this would make it undesirable for the warmer boxes.

The refrigerant in the evaporator coil should be approximately 15°F lower than the space being cooled. If the refrigerant temperature is excessively low, then this will tend to rob the food of its moisture and dry it out. This is particularly applicable to fruits and vegetables. An EPR will elevate the evaporator pressure and thus the refrigerant temperature to bring it more in line with the box temperature.

FUNDAMENTALS OF HVAC/R TEXT REFERENCE
Unit 78

Required Tools and Equipment

Operating refrigeration unit with evaporator pressure regulator	Unit 78
Gauge manifold and temperature sensor	Unit 12

SAFETY REQUIREMENTS

    A. Always read the equipment manual to become familiarized with the refrigeration system and its accessory components prior to startup.

B. Wear safety goggles and gloves when working with refrigerants. Liquid refrigerant can cause frostbite when in contact with eyes and skin.

C. Use low loss hose fittings, or wrap cloth around hose fittings before removing the fittings from a pressurized system or cylinder. Inspect all fittings before attaching hoses.

PROCEDURE

STEP 1. Familiarize yourself with the major components in the refrigeration system including the condenser, compressor, evaporator, and metering device.

A. Evaporator pressure regulators (EPRs) are placed at the outlets of the suction lines for the warmer temperature evaporators. These are adjusted to maintain the desired evaporator pressure and thereby controlling evaporator temperature.

B. A check valve is installed at the outlet of the suction line for the coldest evaporator coil. This prevents migration of the refrigerant from the higher temperature coils to the low temperature coil.

STEP 2. Determine the proper setting for the EPR as follows:

A. You should be able to determine the proper evaporator pressure based upon the desired space temperature. As an example, use the 45°F controlled space temperature in Figure C-5-1. The approximate refrigerant temperature should be 45°F − 15°F = 30°F.

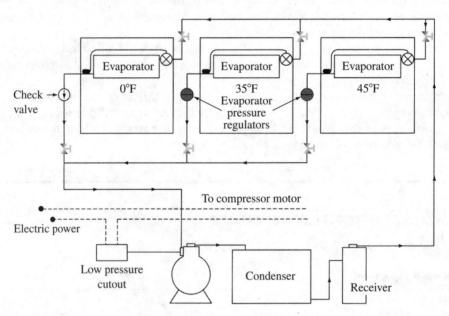

Figure C-5-1

B. You should also consult any operating manual parameters in regard to the expected temperature difference between the space and the evaporator as recommended by the manufacturer. We will be using a 15°F temperature differential for this example, which is most common; however, different types of systems may require other settings.

C. Assuming we are using R-22 then from the P-T chart, Table R-1-1, depicted in Laboratory Worksheet R-1 *Basic Refrigeration System Startup* and from Figure 23-2 in Unit 23 of the *Fundamentals of HVAC/R* text, we can look up the saturated refrigerant temperature.

Inside the evaporator coil, the condition of the refrigerant will always be saturated as long as there is liquid and vapor present together.

From the chart at 30°F, for R-22 the EPR set pressure would be approximately 55 psig. For another type of refrigerant such as R-134a the expected evaporator pressure would be somewhat lower at approximately 26 psig.

Step 3.

A. If there are no pressure gauges currently on the unit then you must connect a gauge manifold to read the evaporator pressure. To help guide you, refer to the procedures in Laboratory Worksheet R-1 *Basic Refrigeration System Startup,* Steps 4, 5, and 6.

B. If you are unable to place temperature probes directly into the refrigerant stream flow, you can use evaporator coil surface temperatures; however, you must adjust for the heat transfer difference through the coil.

Step 4.

A. Start the refrigeration system in the normal cooling mode and allow the system to stabilize and then record the evaporator pressure and temperature below.

_____ psig   _____ °F

B. From your readings, determine what adjustments need to be made if any and set the evaporator pressure regulator by turning the adjusting screw with a screwdriver or Allen wrench dependent on EPR type.

C. Allow the system to run and stabilize prior to shutting down and then carefully disconnect the gauge manifold and any instrumentation that you attached to the unit.

QUESTIONS

(Circle the letter that indicates the correct answer.)

1. The temperature in an evaporator coil is directly related to:
   A. the temperature of the space to be cooled.
   B. the temperature of the condenser.
   C. the pressure in the evaporator.
   D. the compressor pressure.

2. If an evaporator pressure regulator fails shut:
   A. the space will freeze up.
   B. the system will still operate as normal.
   C. the compressor will run continuously.
   D. the space will warm up.

3. An evaporator pressure regulator:
   A. should be used on the warmer evaporators.
   B. should be used on the colder evaporators.
   C. is a safety device.
   D. unloads the compressor at high space temperatures.

4. On a multiple box system, a check valve is usually installed at the outlet of the:
   A. coldest evaporator.
   B. warmest evaporator.
   C. compressor.
   D. condenser.

SETTING A THERMOSTATIC EXPANSION VALVE

LABORATORY OBJECTIVE

The student will demonstrate how to properly set a thermostatic expansion valve to maintain the proper superheat at the evaporator outlet.

LABORATORY NOTES

For this lab exercise there needs to be an operating refrigeration system that has a thermostatic expansion valve. If there are no thermometers or pressure gauges currently on the unit, then you must instrument the refrigeration system using a gauge manifold to measure the evaporator pressure and a thermometer to read the evaporator refrigerant temperature.

If you are unable to place probes directly into the refrigerant stream flow, you can use evaporator coil surface temperatures; however, you must adjust for the heat transfer difference through the coil.

FUNDAMENTALS OF HVAC/R TEXT REFERENCE
Unit 21

Required Tools and Equipment

Operating refrigeration unit with water thermostatic expansion valve	Unit 21
Gauge manifold and temperature sensor	Unit 12

SAFETY REQUIREMENTS

A. Always read the equipment manual to become familiarized with the refrigeration system and its accessory components prior to startup.

B. Wear safety goggles and gloves when working with refrigerants. Liquid refrigerant can cause frostbite when in contact with eyes and skin.

C. Use low loss hose fittings, or wrap cloth around hose fittings before removing the fittings from a pressurized system or cylinder. Inspect all fittings before attaching hoses.

PROCEDURE

STEP 1. Familiarize yourself with the major components in the refrigeration system including the condenser, compressor, evaporator, and metering device.

A. The thermostatic expansion (TEV) supplies the evaporator with enough refrigerant for any and all load conditions. It is NOT a temperature, suction pressure, humidity, or operating control.

B. The thermostatic expansion valve is adjusted to control the refrigerant superheat at the outlet of the evaporator.

C. Many thermostatic expansion valves come preset for 10°F of superheat.

D. Thermostatic expansion valves should not be adjusted unless the evaporator conditions (temperature and pressure) can be measured.

STEP 2. Determine the proper setting for the TEV as follows:

A. The flow of the refrigerant through the TEV is controlled by three pressures: the evaporator pressure, the spring pressure acting on the bottom of the diaphragm, and the bulb pressure opposing these two pressures and acting on the top of the diaphragm.

B. In Figure C-6-1, these three pressures are illustrated. When the three forces are balanced and the valve is in equilibrium as shown, then there should be 10°F of superheat at the evaporator outlet.

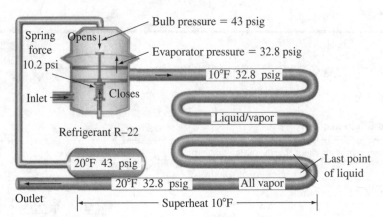

Figure C-6-1

C. The saturation temperature for R-22 at 32.8 psig is 10°F. If the refrigerant pressure is 32.8 psig and the temperature is 20°F, then there would be what is considered as 10°F of superheat.

D. If the superheat temperature decreases, there will be a corresponding decrease in bulb pressure as the TEV would be allowing too much refrigerant to flow. The balance of the three pressures would be disrupted and the valve would begin to close until the pressures arrived at a new equilibrium point.

E. To determine the superheat value, you will need to measure both the pressure and the temperature at the evaporator outlet.

Assuming we are using R-22 and the evaporator pressure is 32.8 psig as shown above then from the P-T chart, Table R-1-1, depicted in Laboratory Worksheet R-1 *Basic Refrigeration System Startup* and from Figure 23-2 in Unit 23 of the *Fundamentals of HVAC/R* text, we can look up the saturated refrigerant temperature.

From the chart at 32.8 psig, for R-22 the saturated temperature would be 10°F. You would then subtract this saturated temperature from the actual measured temperature from the thermometer to determine the total degrees of superheat.

STEP 3.

A. If there are no pressure gauges currently on the unit then you must connect a gauge manifold to read the evaporator pressure. To help guide you, refer to the procedures in Lab R-1 *Basic Refrigeration System Startup*, Steps 4, 5, and 6.

B. If you are unable to place temperature probes directly into the refrigerant stream flow, you can use evaporator coil surface temperatures; however, you must adjust for the heat transfer difference through the coil.

STEP 4.

A. Start the refrigeration system in the normal cooling mode and allow the system to stabilize and then record the evaporator pressure and temperature below.

Measured _____ psig   Measured _____ °F

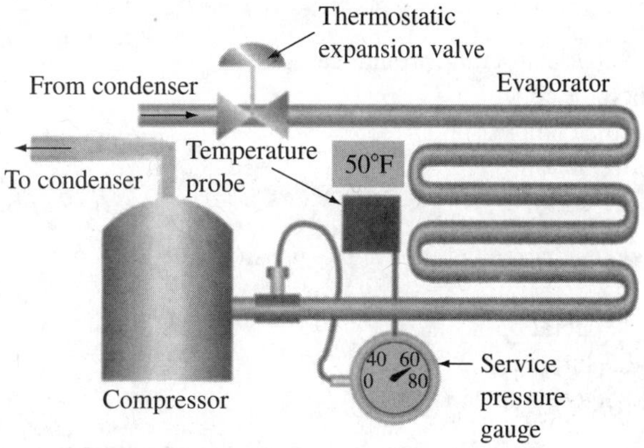

To measure superheat:
1. Find suction pressure
2. Find matching saturation temperature
3. Read temperature leaving evaporator
4. Superheat = temperature leaving − saturation temperature

Figure C-6-2

B. From the saturated pressure-temperature chart, determine the saturated temperature at the measured pressure reading.

C. Calculate the superheat.

Superheat = measured temperature − saturation temperature

D. Determine what adjustments need to be made, if any, and turn the adjusting screw in small increments only. This will change the spring tension, which in turn will change the superheat setting.

E. On many valves, the adjustment is clockwise to increase the superheat and counterclockwise to decrease the superheat. Valve instructions should be checked to be sure of correct adjustments.

F. Allow the system to run and then recheck the superheat once again before making any further changes. It may take some time before the system stabilizes and the final adjustment is complete.

STEP 5.

    A. Continue making any necessary adjustments until the thermostatic expansion valve superheat is correctly set.

    B. Allow the system to run and stabilize prior to shutting down and then carefully disconnect the gauge manifold and any instrumentation that you attached to the unit.

QUESTIONS

To get help in answering some of the following questions, refer to the *Fundamentals of HVAC/R* text Unit 21.

(Circle the letter that indicates the correct answer.)

1. Too high of a TEV superheat setting would lead to:
   A. a flooded evaporator coil.
   B. liquid slugging the compressor.
   C. a starved evaporator coil.
   D. Both A and B are correct.

2. Too low of a TEV superheat setting would lead to:
   A. a flooded evaporator coil.
   B. liquid slugging the compressor.
   C. a starved evaporator coil.
   D. Both A and B are correct.

3. Ice forming on the outside of a thermostatic expansion valve:
   A. indicates blockage.
   B. indicates normal operation.
   C. can crack the diaphragm.
   D. should be chipped off with an ice pick.

4. Raising the temperature of a fluid above its saturation temperature is:
   A. superheating.
   B. subcooling.
   C. sublimation.
   D. saturation absorption.

5. If the sensing bulb for a thermostatic expansion valve becomes detached from the evaporator coil:
   A. the valve will begin to close and starve the evaporator.
   B. the valve will begin to open and starve the evaporator.
   C. the valve will begin to close and flood the evaporator.
   D. the valve will begin to open and flood the evaporator.

6. If the diaphragm on a thermostatic expansion valve fails, the valve:
   A. will begin to close and starve the evaporator.
   B. will begin to open and starve the evaporator.
   C. will begin to close and flood the evaporator.
   D. will begin to open and flood the evaporator.

7. If the evaporator coil is cold at the inlet but is warm about halfway down, then this could indicate:
   A. a high superheat.
   B. a low superheat.
   C. the proper amount of superheat.
   D. floodback.

INSTALLING A REPLACEMENT THERMOSTAT

LABORATORY OBJECTIVE
The purpose of this lab is to demonstrate your ability to install a replacement thermostat.

LABORATORY NOTES
Many thermostats are multiday programmable, and may even be multizone programmable. Older thermostats may have a mercury switch and the proper disposal for this type is at a toxic waste collection site. Check with your instructor regarding the necessary procedures for handling mercury switches.

FUNDAMENTALS OF HVAC/R TEXT REFERENCE
Unit 35

Required Tools and Equipment

Operating HVAC unit	
Toolkit	Unit 10
Multimeter	Unit 11
Clamp-on ammeter	Unit 11
Temperature sensor	Unit 12
Thermostat	Unit 35

SAFETY REQUIREMENTS

A. Check all circuits for voltage before doing any service work.

B. Stand on dry nonconductive surfaces when working on live circuits.

C. Never bypass any electrical protective devices.

PROCEDURE

STEP 1. Collect the HVAC unit data and fill in the chart.

HVAC Unit Data

Unit description	
Fuel type	
Ignition system	
Control transformer	VA rating = _____ Location _____
Fan relay type and number	
Heating circuit device and rated amperage draw	
Cooling capacity	
Number of stages of heat	
Number of stages of cooling	
List other functions to control or supervise	

STEP 2. Collect the thermostat data and fill in the chart.

Thermostat Data

Make:	Model number:
Stages of heat:	Stages of cooling:
Heat anticipator: (circle one) Adjustable Nonadjustable	
Cool anticipator: (circle one) Adjustable Nonadjustable	
Subbase model number:	
Subbase switching (circle all that apply) HEAT OFF AUTO COOL FAN ON FAN AUTO	
List special thermostat features if any:	

STEP 3. Complete the Thermostat Installation Checklist and make sure to check each step off in the appropriate box as you finish it. This will help you to keep track of your progress.

Thermostat Installation Checklist

Step	Procedure	Check
1	Make sure all power to the unit is off. Lock and tag the power panel before removing any parts.	
2	Check existing thermostat wire and size.	
3	Thermostat wire size = _____ (18 gauge is typical)	
4	Number of conductors = _____	
5	Color of wires (circle all that apply). White Red Green Blue Yellow Other	
6	Install and level new thermostat subbase in a stable location.	
7	Match and record thermostat wire color with subbase terminals. W1_____ W2_____ O_____ B_____ R or RH_____ RY_____ G_____ Y1_____ Y2_____	
8	Install a jumper from RH to W1.	
9	Turn the power to the unit on and observe the heat operation.	
10	Measure the amperage draw of the heating circuit. Amperage =_____	
11	Turn the power to the unit off. Lock and tag the power panel before removing any parts.	
12	Install the thermostat on the subbase.	
13	Set the heat anticipator to the measured amperage draw.	

STEP 4. After installing the thermostat you are now ready to check for proper operation. You will follow the same procedure of working through a checklist. This will help you to keep track of your progress. Complete the Operational Checklist and make sure to check each step off in the appropriate box as you finish it.

Operational Checklist

Step	Procedure	Check
1	Turn the thermostat to the OFF position.	
2	Obtain an accurate room temperature at the thermostat location. _____	
3	Turn the setpoint setting until the contacts open.	
4	Compare the setpoint with the room temperature. Setpoint _____ Room temperature _____	
5	Calibrate setpoint to equal room temperature.	
6	Turn the power ON.	
7	Start and set clock if required.	
8	Use the thermostat instructions to program the thermostat if programmable.	
9	Turn HEAT-OFF-COOL switch to the HEAT position and turn setpoint to above room temperature.	
10	Observe heat operation.	
11	Turn fan switch to ON position. Observe fan turn on.	
12	Turn fan switch to AUTO position. Observe fan turn off.	
13	Turn HEAT-OFF-COOL switch to COOL position.	
14	Turn setpoint to below room temperature.	
15	Observe cooling operation.	
16	Turn setpoint to above room temperature.	
17	Observe cooling off.	
18	Install thermostat cover.	

PROGRAM THERMOSTAT

LABORATORY OBJECTIVE
The purpose of this lab is to demonstrate your ability to program a thermostat.

LABORATORY NOTES
Energy conservation is important for heating and air conditioning systems. The electronic programmable clock thermostat is a popular energy management tool for the residential and light commercial market. This device allows the customer to automatically change the setpoint temperatures for occupied times and set back or set up the unoccupied temperature depending on the heating or cooling mode.

Since each programmable thermostat operates differently, keeping the instruction booklet is essential. Certain features have common functions but the function name may vary from brand to brand. Some are full seven day programmable while others are five-day/two-day clocks with all weekdays treated the same and both weekend days treated the same. A business would usually require a full seven-day programmable thermostat. Typically there are a maximum of four daily times (wake, leave, come home, and sleep). The four times are designed for two setback periods; one during the night and one during the day when people are at work. This type of thermostat can be used for business applications by setting the midday setback at one temperature all day or stacking the middle time points on one point in time.

FUNDAMENTALS OF HVAC/R TEXT REFERENCE
Unit 35

Required Tools and Equipment

Programmable thermostat	Unit 35

SAFETY REQUIREMENTS
Check all circuits for voltage before doing any service work.

PROCEDURE

STEP 1. Obtain a five-day/two-day programmable thermostat with four daily setback periods. Complete the Thermostat Programming Checklist and make sure to check each step off in the appropriate box as you finish it. This will help you to keep track of your progress.

Thermostat Programming Checklist (five-day/two-day)

Step	Procedure	Check
1	Read and follow the manufacturer's instructions.	
2	Set the clock to the correct time of day.	
3	Set the five-day (weekday) setting for: 6:00 AM (wake)—set to 70°F 7:30 AM (leave)—set to 62°F 4:30 PM (home)—set to 70°F 10:00 PM (sleep)—set to 62°F	
4	Set the two-day (weekend) setting for: 8:00 AM (wake)—set to 70°F 11:00 AM (leave)—set to 65°F 5:00 PM (home)—set to 70°F 11:30 PM (sleep)—set to 60°F	

STEP 2. Obtain a seven-day programmable thermostat. Complete the Thermostat Programming Checklist and make sure to check each step off in the appropriate box as you finish it.

Thermostat Programming Checklist (seven-day)

Step	Procedure	Check
1	Read and follow the manufacturer's instructions.	
2	Set the clock to the correct time of day.	
3	Set the seven-day setting for a typical retail store: 9:00 AM to 8:00 PM Monday through Thursday—set to 70°F 9:00 AM to 11 PM Friday—set to 70°F 8:00 AM to 11:30 PM Saturday—set to 70°F 9:00 AM to 11 PM Sunday—set to 70°F Provide a 60°F temperature for the unoccupied time.	
4	Allow lead time for morning warm up based on a 5°F/hr pick-up. This means that at 8:00 AM every day except for Saturday, which would be 7:00 AM, the temperature would be set for 65°F. The thermostat could also be set to allow for the unoccupied setting temperature to begin 1/2 hour before store closing, and this would still allow for a comfortable temperature and reduce energy usage.	

ELECTRICAL MULTIMETERS

LABORATORY OBJECTIVE

The student will demonstrate how to properly use an electrical multimeter.

LABORATORY NOTES

For this lab exercise there should be a typical multimeter and a live electrical circuit that can be used for test purposes.

Testing electrical circuits is an important skill that each technician needs to develop. Although practice and experience are significant, a high degree of success is obtainable by following a proven procedure such as the following:

A. Know the unit electronically. This means understanding the proper function of each control and the sequence of the control operation.

B. Be able to read schematic wiring diagrams and have them available.

C. Be able to use the proper electrical test instruments. Know the instrument. Read instructions carefully before using.

FUNDAMENTALS OF HVAC/R TEXT REFERENCE

Unit 11

Required Tools and Equipment

Digital and analog multimeters	Unit 11
Clamp-on ammeter	Unit 11
Live electrical circuit and resistor	

SAFETY REQUIREMENTS

A. Check all circuits for voltage before doing any service work.

B. Stand on dry nonconductive surfaces when working on live circuits.

C. Never bypass any electrical protective devices.

PROCEDURE

STEP 1. Familiarize yourself with the operation of an analog multimeter by checking the resistance of a known resistor. See Figure E-1-1.

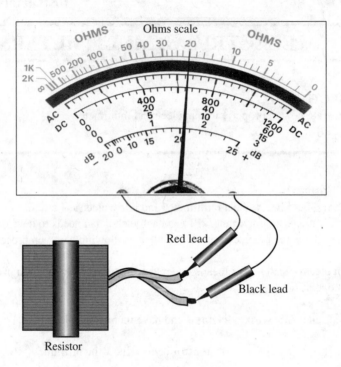

Figure E-1-1

A. The ohmmeter uses a battery to furnish the current needed for resistance measurements. The higher the resistance, the lower the current flow.

B. Whenever checking resistance, the power to the circuit being tested must be shut off.

C. Set the meter to measure resistance (Ω).

D. Zero the meter by touching the red and black leads together. Since they are touching, the resistance should read zero. When they are apart, the resistance should be infinite.

E. Touch the red lead to one end of the resistor and the black lead to the other end and measure the resistance. Record the reading.

Resistance = _____ Ω

F. Repeat the same measurement using a digital multimeter and record the reading. How do the two readings compare?

Resistance = _____ Ω

STEP 2. Check the voltage of a circuit using a multimeter. See Figure E-1-2.

A. Voltmeters are connected in parallel with the load to read the voltage drop. A knob in the center face of the meter adjusts the meter to the scale being used. Always start to measure voltage using the highest range on the meter.

240

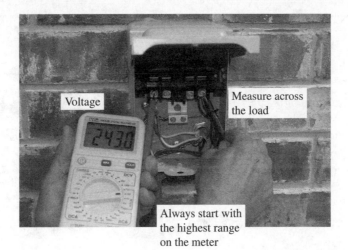

Figure E-1-2

B. Examine the wiring diagram for the circuit to be tested and determine how the leads from the meter should be connected.

C. Once familiar with circuit, turn on the power and measure the voltage. Record the reading.

Voltage = _____ V

STEP 3. Check the amperage of a circuit using a clamp-on ammeter. See Figure E-1-3.

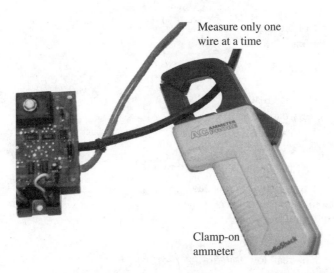

Figure E-1-3

A. When taking a reading, always start with the highest possible scale and then work down to the most appropriate scale.

B. Never put the clamp around two different wires at the same time.

C. Examine the wiring diagram for the circuit to be tested and determine where to clamp the ammeter.

D. Once familiar with the circuit, turn on the power and measure the amperage. Record the reading.

Amperage = _____ A

QUESTIONS

To get help in answering some of the following questions, refer to the *Fundamentals of HVAC/R* text Unit 11.

(Circle the letter that indicates the correct answer.)

1. When measuring voltage:
   A. attach the lead in series with the load.
   B. always use the highest scale on the meter first.
   C. always use the lowest scale on the meter first.
   D. use a clamp-on ammeter.

2. Electrical potential is measured with a/an:
   A. ohmmeter.
   B. wattmeter.
   C. megohm meter.
   D. voltmeter.

3. Always remember that when using an ohmmeter:
   A. start with the lowest scale reading first.
   B. make sure that the circuit is deenergized.
   C. make sure that the circuit is energized.
   D. the dial must be tapped gently to calibrate the needle.

4. A clamp-on ammeter can be used to measure current flow through:
   A. a single wire only.
   B. multiple wires.
   C. multiple wires of different polarity.
   D. motor foundations.

5. Using an ohmmeter on a live circuit:
   A. is always correct.
   B. will provide a voltage reading.
   C. must be done very carefully.
   D. may destroy the meter.

6. When the two leads of an ohmmeter are touched together:
   A. there will be a spark.
   B. there will be zero resistance indicated.
   C. there will be infinite resistance indicated.
   D. The meter will short out.

7. Most clamp-on ammeters:
   A. cannot accurately read low amp draws.
   B. can accurately read low amp draws.
   C. will measure the resistance of the wire along with the current.
   D. will measure the voltage of the wire along with the current.

8. Digital meters:
   A. are less accurate than analog meters.
   B. are generally cheaper than comparable analog meters.
   C. are more easily damaged than analog meters.
   D. can be accurate to three decimal places.

9. When checking the voltage of a DC circuit:
   A. only one lead from the meter is required.
   B. verify the correct polarity of the probes that are to be used before connecting the meter to the circuit.
   C. start with the lowest scale.
   D. All of the above are correct.

10. When measuring the resistance of a circuit, the reading is 0 Ω, therefore:
   A. this indicates an open circuit.
   B. this means that the meter is faulty.
   C. this indicates a short circuit.
   D. Any of the above could be correct.

11. Solid state circuits:
   A. may be tested with any type of ohmmeter.
   B. may only be tested with an analog ohmmeter.
   C. may only be tested with a digital ohmmeter.
   D. should normally not be tested with an ohmmeter.

12. Voltage is normally measured:
   A. in series with the load.
   B. across the load.
   C. Both A and B are correct.
   D. None of the above is correct.

ELECTRICAL SOLDERING

LABORATORY OBJECTIVE
The student will demonstrate how to properly solder an electrical wiring connection.

LABORATORY NOTES
For this lab exercise there should be a component in a circuit that needs to be soldered into place. It is preferable that the component have soldering lugs attached for wire placement. If it is a working circuit, then it should be tested for proper operation after the soldering exercise has been completed.

FUNDAMENTALS OF HVAC/R TEXT REFERENCE:
Unit 16

Required Tools and Equipment

Electrical soldering iron or gun	Unit 16
Wire, solder, component terminal lugs	Unit 16

SAFETY REQUIREMENTS

    A. The soldering iron is extremely hot and can burn skin, clothes, and tabletops if not handled carefully.

    B. A soldering iron stand should be used to hold the soldering iron at a safe distance above the work surface when warming up and cooling down.

PROCEDURE

STEP 1. Prepare the soldering iron for use.

    A. Most electrical soldering uses a rosin flux-cored wire. Rosin is inactive at low temperatures and will not corrode the electrical parts even if the rosin is left on following soldering.

B. The tip of an electric soldering iron or gun is made of copper. A new tip or a damaged tip must first be tinned with solder before it can be used. See Figure E-2-1.

C. First clean the tip surface with a sand cloth or file.

D. Apply soldering flux and then turn on the gun.

E. Press the solder against the surface until the solder begins to melt.

F. Turn off the gun and rub the tip surface with the solder until the entire surface has been tinned with solder.

Fluxing soldering gun tip in preparation for tinning

Tip

Rosin flux

Figure E-2-1

STEP 2. Attach the wires to the component.

A. Thread the wire through the eyelet of the component terminal lug as shown in Figure E-2-2. Make a solid mechanical connection by bending the wire around in a tight loop.

B. Turn on the soldering gun and allow the solder on the tip of the gun to melt.

C. Hold this molten solder against the electrical wire ends.

Figure E-2-2

D. Place the solder wire against the opposite side of the wires being joined as shown in Figure E-2-3.

E. When the wires reach soldering temperature, the solder will melt and flow into the joint. Gently remove the soldering tip and solder wire from the connection. Allow the solder to cool and solidify completely before the solder joint is moved.

Figure E-2-3

QUESTIONS

To get help in answering some of the following questions, refer to the *Fundamentals of HVAC/R* text Unit 16.

(Circle the letter that indicates the correct answer.)

1. Acid core solder:
    A. is always used for electrical component soldering.
    B. should never be used for electrical component soldering.
    C. can be used in place of rosin core solder for electrical connections only.
    D. is nonreactive to electrical components.

2. In regard to soldering electrical connections:
    A. allow the gun or soldering iron to heat up before putting it on the connection.
    B. the gun should never be heated in advance.
    C. always use acid core solder.
    D. flex the joint before it cools to ensure that it is tight.

3. Tinning the tip of a soldering iron:
    A. is an unnecessary step.
    B. is accomplished by rolling the tip of the soldering iron on a flat aluminum cookie sheet.
    C. is good practice.
    D. means to file down the outside dimension of the tip.

4. The proper way to apply heat for soldering is to:
  A. allow the solder to melt on the tip of the soldering iron and drip on to the joint.
  B. heat the component at its opposite end.
  C. hold the soldering iron at least 3 in from the work.
  D. heat the electrical wire to a temperature at which the solder will melt.

5. When soldering electric circuit boards:
  A. excessive heat is never a problem.
  B. excessive heat should be drawn off with a special heat sink.
  C. excess solder should always be applied.
  D. never preheat the soldering iron.

6. The tip of an electric soldering iron is usually made from:
  A. stainless steel.
  B. lead.
  C. pig iron.
  D. copper.

7. A problem with using acid core solder is that:
  A. it is difficult to remove all the acid from within the fine wires of the electrical connection.
  B. acid is reactive to electrical components.
  C. acid will damage electrical components.
  D. All of the above are correct.

8. Before tinning, a soldering iron tip should be sanded and fluxed:
  A. True.
  B. False.

TESTING A CAPACITOR

LABORATORY OBJECTIVE

The student will demonstrate how to properly discharge and test a capacitor used on a motor starting circuit.

LABORATORY NOTES

Many single phase motors require capacitors. Capacitors can be dangerous. Capacitors can hold a high voltage charge even after the power is turned off. Always discharge capacitors before touching them.

FUNDAMENTALS OF HVAC/R TEXT REFERENCE

Unit 30

Required Tools and Equipment

Capacitor	Unit 30
Ohmmeter	Unit 11
20,000 Ω, 2 W resistor	Unit 30

SAFETY REQUIREMENTS

A. Turn the power off if the capacitor to be tested is installed in an operating system. Lock and tag out the power supply.

B. Confirm the power is secured by testing for zero voltage with your meter.

C. Capacitors should be discharged with a bleed resistor.

D. Never discharge a capacitor with a screwdriver or other metal device touching the terminals. The spark that is generated externally can also occur internally, damaging the compressor. See Figure E-3-1.

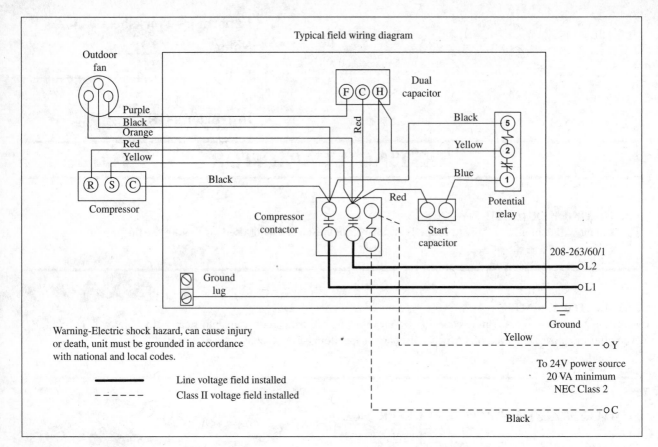

Figure E-3-1

PROCEDURE

STEP 1. Discharge the capacitor as follows:

A. A Place the capacitor in a protective case and connect a 20,000 Ω, 2 W resistor across the terminals. See Figure E-3-2.

B. Most start capacitors have a bleed resistor; however, it is good practice to make sure the charge has been bled off.

C. To test the capacitor, disconnect it from the wiring and place the ohmmeter leads on the terminals as shown in Figure E-3-3. The ohmmeter should have at least an R × 100 scale. Familiarize yourself with electrical meters (refer to Laboratory Worksheet E-1, *Electrical Meters*).

D. If the capacitor is good, the needle will make a rapid swing toward zero and slowly return to infinity as shown in Figure E-3-4.

E. If the capacitor has an internal short, the needle will stay at zero, indicating the instrument will not take a charge.

F. In replacing a capacitor, it is desirable to use an exact replacement—a capacitor with the same mfd rating and voltage limit rating.

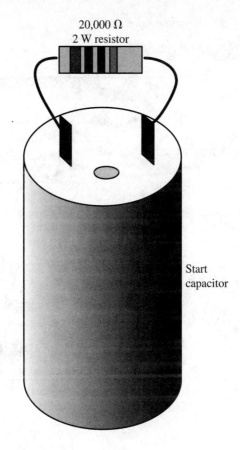

20,000 Ω
2 W resistor

Start
capacitor

Figure E-3-2

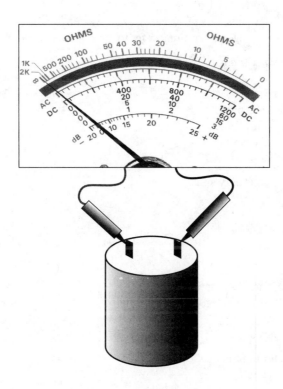

Figure E-3-3

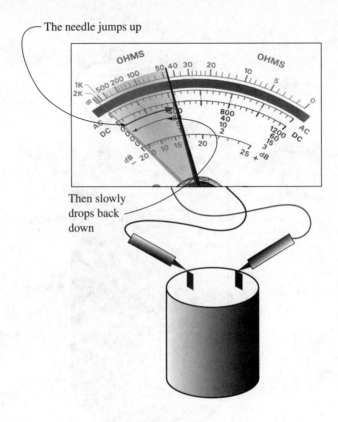

The needle jumps up

Then slowly
drops back
down

Figure E-3-4

QUESTIONS

To get help in answering some of the following questions, refer to the *Fundamentals of HVAC/R* text, Unit 30.

(Circle the letter that indicates the correct answer.)

1. Start capacitors:
   A. are high capacity (100–800 mfd).
   B. are low capacity (2–40 mfd).
   C. are only charged when the circuit is live.
   D. can be discharged with a screwdriver.

2. Capacitors rated for continuous duty are:
   A. start capacitors.
   B. run capacitors.
   C. Both A and B are correct.
   D. None of the above is correct.

3. In regard to capacitors:
   A. start and run capacitors are never interchangeable.
   B. start capacitors are always rated for continuous duty.
   C. start capacitors are low capacity (2–40 mfd).
   D. All of the above are correct.

4. A red dot on a run capacitor terminal indicates:
   A. the high side of the system.
   B. the free end.

C. it should be connected to the run terminal.
D. its mfd rating.

5. If a capacitor has a bleed resistor:
   A. it must be removed before testing.
   B. the capacitor can still be tested with the bleed resistor in place.
   C. it will fail open.
   D. Both B and C are correct.

6. When testing a capacitor:
   A. the ohmmeter should have at least an R $\times$ 100 scale.
   B. the ohmmeter should have at least an R $\times$ 1000 scale.
   C. the ohmmeter should have at least an R $\times$ 10 scale.
   D. None of the above is correct.

7. Discharging a capacitor by shorting its terminals:
   A. is always the preferred method.
   B. will cause violent sparking.
   C. will melt the tip of a screwdriver.
   D. will never damage the compressor motor.

8. If the capacitor is good, the ohmmeter needle will stay at zero:
   A. True.
   B. False.

MOTOR INSULATION TESTING

LABORATORY OBJECTIVE

The student will demonstrate how to properly test for leakage resistance from motor windings.

LABORATORY NOTES

The wiring that is used for motor windings is insulated with a thin coat of varnish. As the motor is used and the windings are heated, this material will slowly carbonize. As it carbonizes, it will slightly change to a darker tan and eventually on to black. When it is completely carbonized, it does not provide any insulating capacity. See Figure E-5-1.

An instrument referred to as a megger tests the degree to which carbonization has taken place within the motor winding insulation.

FUNDAMENTALS OF HVAC/R TEXT REFERENCE
Unit 30

Required Tools and Equipment

Motor	Unit 30
Megger	Unit 11

SAFETY REQUIREMENTS

A. Turn the power off if the winding to be tested is installed in an operating system. Lock and tag out the power supply.

B. Confirm the power is secured by testing for 0 voltage with a multimeter.

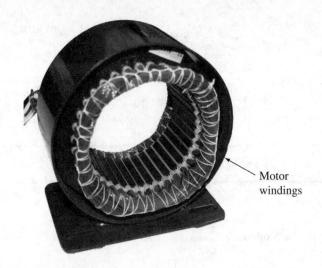

Figure E-5-1

PROCEDURE

STEP 1. Familiarize yourself with the operation of a megger. See Figure E-5-2.

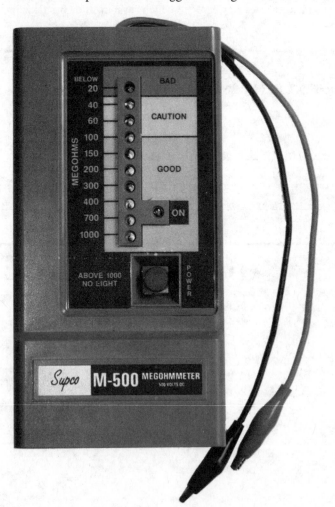

Figure E-5-2

A. Meggers can test leakage at high voltage. High voltage can be thought of as 500 V for 208 to 240 V motors and 1,000 V for 480 V motors.

B. Meggers may be battery operated or have a built in hand cranked generator.

C. They may detect insulation faults, while an ordinary multimeter using a few volts DC would show a satisfactory reading.

D. The megger shown in Figure E-5-2 has a scale that reads from 20 to 1,000 megohms.

E. Below 30 megohms would be considered bad insulation.

F. 30 to 100 megohms would indicate that the windings are questionable.

G. 100 megohms or greater would be considered good.

STEP 2. Prepare to test the motor insulation as follows:

A. Turn the power off if the winding to be tested is installed in an operating system. Lock and tag out the power supply.

B. Confirm the power is secured by testing for 0 voltage with a multimeter.

C. Attach one lead of the megger to the motor terminal and the other lead to ground (the motor or compressor frame is suitable).

D. Record your reading below.

Run winding _____Ohms

Start winding _____Ohms

Common terminal _____Ohms

E. Larger motors are checked for insulation breakdown by having megger readings taken on a regular basis in accordance with the preventative maintenance schedule. This provides a record of the motor condition over time and a lowering resistance indicates that the windings are becoming increasingly carbonized.

TESTING START RELAYS

LABORATORY OBJECTIVE
The student will demonstrate how to properly test a starting relay.

LABORATORY NOTES
A starting relay assists in starting the motor by allowing current flow to the starting winding through the start capacitor and the normally closed contacts in the relay. See Figure E-6-1.

When the motor comes up to speed the start relay coil energizes and acting as an electromagnet it will open the normally closed contacts in the start relay. See Figure E-6-2.

FUNDAMENTALS OF HVAC/R TEXT REFERENCE
Units 29 and 30

Required Tools and Equipment

Motor control circuit	Unit 30
Multimeter	Unit 11

SAFETY REQUIREMENTS

A. Turn the power off if the winding to be tested is installed in an operating system. Lock and tag out the power supply.

B. Confirm that the power is secured by testing for 0 voltage with a multimeter.

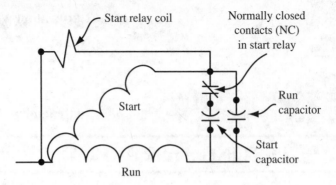

Figure E-6-1

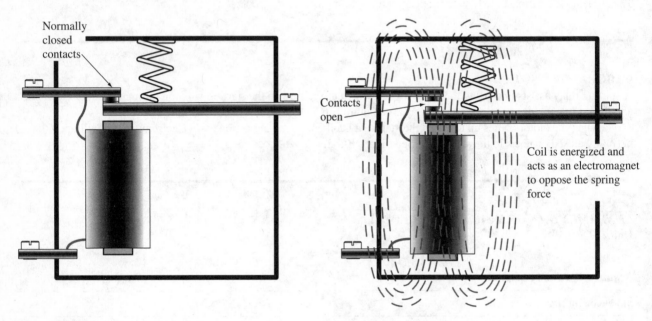

Figure E-6-2

PROCEDURE

STEP 1. Familiarize yourself with electrical meters (refer to Laboratory Worksheet E-1, *Electrical Meters*).

A. Many start relay coils have higher resistance than average control circuit relays.

B. Be sure to test the coil on the R × 100 scale before deciding that the relay is defective.

C. The pull-in/drop-out voltage of the relay is unique. Do not attempt to replace it with an ordinary, similar voltage relay.

STEP 2. Test the relay coil as follows (see Figure E-6-3):

    A. Start with the resistance reading on the highest scale.

    B. Record the resistance reading. _____Ohms

    C. Compare this reading to the manufacturer's data to see if the coil is satisfactory.

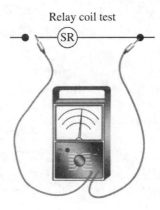

Relay coil test

Figure E-6-3

STEP 3. Test the relay contacts as follows (see Figure E-6-4):

    A. When testing start relay contacts, the contacts should be closed and the ohmmeter will read zero resistance.

Relay contact test

Figure E-6-4

    B. Sometimes while in operation the contacts stick closed and become badly burned. This will make for a poor connection, which will be indicated by a positive resistance reading.

    C. If such is the case, replace the relay. Do not attempt to clean the contacts.

    D. Always use the identical replacement. An improper substitution can damage the motor.

    E. The replacement must be mounted in the same position as the original and connected in the same way.

QUESTIONS

To get help in answering some of the following questions, refer to the *Fundamentals of HVAC/R* text Unit 30.

(Circle the letter that indicates the correct answer.)

1. When replacing a start relay:
   A. always use an identical replacement.
   B. do not attempt to replace it with an ordinary, similar voltage relay.
   C. remember that an improper substitution can damage the motor.
   D. All of the above are correct.

2. Burned start relay contacts:
   A. should be cleaned with a fine grit emery cloth.
   B. should be carefully filed to remove any pitting.
   C. should be cleaned with the proper solvent.
   D. should be replaced.

3. Good start relay contacts should have:
   A. an infinite resistance when closed.
   B. a fairly high resistance when closed.
   C. no resistance when closed.
   D. a resistance of exactly 100 megohms.

4. Start relay coils have lower resistance than average control circuit relays:
   A. True.
   B. False.

5. If the start relay contacts stick closed while the motor is running:
   A. the motor will overspeed.
   B. the motor will stall.
   C. the start winding could be damaged.
   D. Both A and C are correct.

6. An infinite resistance in a start relay coil:
   A. indicates a short.
   B. indicates an open.
   C. is normal.
   D. None of the above is correct.

7. The pull-in/drop-out voltage:
   A. is unique to each start relay.
   B. is always the same for all start relays.
   C. is easily adjusted.
   D. is directly related to the frequency of the circuit.

WIRING DIAGRAMS

LABORATORY OBJECTIVE

The student will demonstrate how to properly draw a wiring schematic from a connection diagram.

LABORATORY NOTES

The student may draw the wiring schematic from the connection diagram in Figure E-7-1 or from an alternate connection diagram supplied by the Lab Instructor. The connection diagram shown in Figure E-7-1 is for an air cooled condensing unit. The control system consists of a wiring panel enclosing the compressor starter, the start relay, the run capacitor, a thermostat and switch combination, a start capacitor, and a junction box.

External to the wiring panel are the fan motor, power supply, compressor motor, junction box, and high/low pressure control.

FUNDAMENTALS OF HVAC/R TEXT REFERENCE

Unit 31

Required Tools and Equipment

None	

SAFETY REQUIREMENTS

None

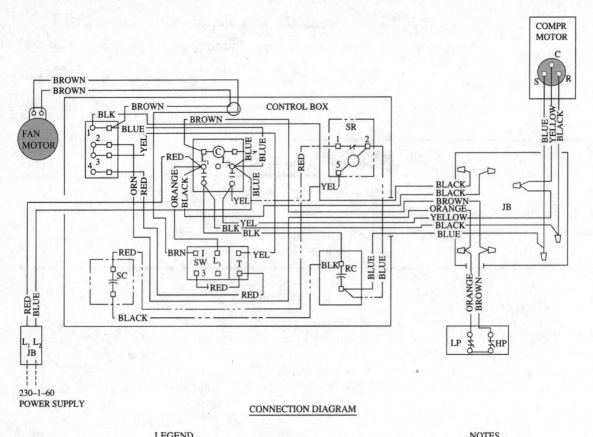

CONNECTION DIAGRAM

LEGEND

C	CONTACTOR	S	START
RC	RUN CAPACITOR	R	RUN
SC	START CAPACITOR	C	COMMON
SR	START RELAY		
T	THERMOSTAT	- - - -	FIELD WIRING
SW	SWITCH	———	FACTORY WIRING
HP	HIGH-PRESSURE SWITCH	- - -	ALTERNATE CSR WIRING
LP	LOW-PRESSURE SWITCH		
JB	JUNCTION BOX		

NOTES

1. FAN MOTOR PROVIDED WITH INHERENT THERMAL PROTECTOR.

2. COMPR. MOTOR PROVIDED WITH INHERENT OVERLOAD PROTECTOR.

3. MAX. FUSE SIZE 30-AMP DUAL ELEMENT.

Figure E-7-1

PROCEDURE

STEP 1. The first step in preparing the schematic is to locate the loads and determine the number of circuits. There are five loads and therefore five circuits.

    A. The five loads are:

        1. _____

        2. _____

        3. _____

        4. _____

        5. _____

STEP 2. The second step in preparing the schematic is to locate the switches in each circuit.

A. The switches in each circuit are:

1. _____

2. _____

3. _____

4. _____

5. _____

STEP 3. The third step in preparing the schematic is to draw each circuit.

A. Draw the compressor *motor* circuit and start relay coil circuit in the space provided below.

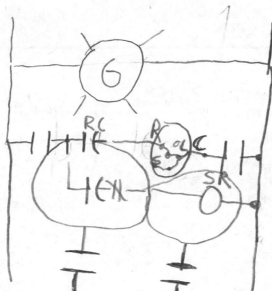

B. Draw the fan *motor* circuit, the compressor contactor circuit, and the green test lamp circuit in the space provided below.

C. Combine parts A and B together into one complete schematic wiring diagram in the space provided below.

QUESTIONS

To get help in answering some of the following questions, refer to the *Fundamentals of HVAC/R* text Unit 31.

(Circle the letter that indicates the correct answer.)

1. The three types of wiring diagrams are:
   A. internal, external, and connection.
   B. external, connection, and schematic.
   C. internal, connection, and schematic.
   D. external, connection, and plan.

2. The electric diagram symbol shown below is a:

   A. thermal relay.
   B. thermal cutout.
   C. flow switch.
   D. closed contact.

3. The electric diagram symbol shown below is a:

   A. thermal relay.
   B. transformer.
   C. flow switch.
   D. closed contact.

4. The electric diagram symbol shown below is a:

A. thermal relay.
B. transformer.
C. flow switch.
D. closed contact.

5. The electric diagram symbol shown below is a:

A. motor winding.
B. transformer.
C. flow switch.
D. closed contact.

6. The electric diagram symbol shown below will:

A. close on rising temperature.
B. open on rising temperature.
C. close on rising pressure.
D. open on rising pressure.

7. The electric diagram symbol shown below will open on rising temperature:

Temperature
Control

A. True.
B. False.

SOLDERING COPPER PIPE

LABORATORY OBJECTIVE

The student will demonstrate how to properly solder a copper pipe connection.

LABORATORY NOTES

For this lab exercise a fitting will be soldered to copper piping. The fitting will then be tested to determine if it is satisfactory.

Soldering takes place below 840°F and is used for water or condensate drains. *Solder is not approved for refrigerant line joints.*

Soldering to join copper pipes uses an air acetylene torch, air MAPP torch, or air propane torch. Oxyacetylene is not recommended for soldering due to high flame temperatures. See Table M-1-1.

FUNDAMENTALS OF HVAC/R TEXT REFERENCE

Unit 16

Required Tools and Equipment

Air acetylene or air propane/MAPP torch	Unit 16
Copper piping, fitting, solder	Units 15 & 16
Hacksaw, pliers, and flat screwdriver	Unit 9

SAFETY REQUIREMENTS

A. One hundred percent cotton or leather clothing is the best material to wear while brazing, soldering, or welding.

B. Shirts should have long sleeves and work gloves must be worn (cloth, leather palm, or all leather).

C. Lead solder should be avoided.

PROCEDURE

STEP 1. Prepare the copper piping and fitting for use.

   A. Use an abrasive sanding cloth (Figure M-1-1) or wire brush (Figure M-1-2) to remove all contaminants from the surface of the pipe.

TABLE M-1-1 Common soldering and brazing metal and fluxes showing base metals that can be joined

	Alloy	Flux type	Base metal
S O L D E R	95-5 Tin antimony solder[1]	C-Flux	Copper pipe, brass, steel
	95-5 Tin antimony solder	Rosin	Copper pipe, copper wiring, brass
	95-5 Tin antimony solder	Acid	Copper pipe, brass, steel, galvanized sheet metal
	98-2 Tin silver solder	Mineral based flux	Copper pipe, brass, steel
	40-60 Cadmium zinc solder	Specific flux from solder manufacturer	Aluminium
B R A Z I N G	Copper phosphorus silver brazing BCuP	1% to 15% Silver no flux required	Copper pipe
	Copper phosphorus silver brazing BCuP	1% to 15% Silver mineral based flux	Copper pipe to brass, brass, steel
	Copper silver brazing BCuP	45% Silver mineral flux	Copper pipe to steel, brass, steel

[1]The percentages of the materials in the flux are given in the numbers, for example 95% tin, 5% antimony.

Figure M-1-1

Figure M-1-2

B. Do not touch the cleaned surfaces with your hands as oil from your fingers can prevent solder from flowing completely into a joint.

C. Use a brush to apply the flux to the end of the pipe.

D. *Do not apply flux to the very end of the pipe.*

E. Apply the flux to approximately 1/16 in to 1/8 in from the end of the pipe to avoid flux contamination into the system.

STEP 2. Once the joint is prepared, then continue as follows:

A. Three quarters of an inch of solder is adequate to make soldered joints in 1/2 in through 1 in diameter copper pipe. Bend the solder approximately 3/4 in from the end. See Figure M-1-3.

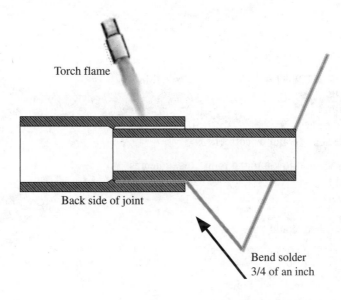

Torch flame

Back side of joint

Bend solder
3/4 of an inch

Figure M-1-3

B. Use an air fuel torch and begin by heating the pipe.

C. As the pipe begins to be heated, move the torch onto the fitting and pipe.

D. Periodically test the back side of the joint with the tip of the solder.

E. Once the solder begins to melt, remove the torch.

F. Continue adding the solder until the joint is filled.

G. Ideally a small fillet of solder should be left at the joint surface.

H. Pushing more solder into the joint simply results in solder bb's being formed inside the piping.

STEP 3. Destructively test the soldered joint as follows:

A. Once cooled, use a hacksaw to slice a 45° angle through the fitting but not the pipe.

B. Place a flat bladed screwdriver into the slot and twist to release the pipe fitting from the copper pipe.

C. Using a pair of pliers, peel back the copper fitting to expose the soldered surface.

D. If the surface is smooth and has no large voids, the soldering was successful.

QUESTIONS

To get help in answering some of the following questions, refer to the *Fundamentals of HVAC/R* text Unit 16.

(Circle the letter that indicates the correct answer.)

1. Some of the common problems with soldering include:
   A. overheating.
   B. underheating.
   C. Both A and B are correct.
   D. None of the above is correct.

2. Tiny bubbles in the solder indicate:
   A. overheating.
   B. underheating.
   C. the wrong solder type.
   D. Any of the above could be correct.

3. If the solder does not flow into the joint, this indicates:
   A. overheating.
   B. underheating.
   C. the wrong solder type.
   D. Any of the above could be correct.

4. Soldering flux should be applied:
   A. with a brush.
   B. 1/8 of an inch from the end of the pipe.
   C. so that it is spread around the joint.
   D. All of the above are correct.

5. The solder:
   A. should be introduced to the back side of the pipe.
   B. should be bent approximately 3/4 in from the end.
   C. is usually a tin and antimony base.
   D. All of the above are correct.

BRAZING

LABORATORY OBJECTIVE

The student will demonstrate how to properly braze a copper pipe connection.

LABORATORY NOTES

For this lab exercise a fitting will be brazed to copper piping. The fitting will then be tested to determine if it is satisfactory.

Brazing takes place at temperatures above 840°F and codes require that refrigerant joints be made using silver brazing alloys.

The most commonly used torch for this type of brazing is the air acetylene torch. An air acetylene flame burns at approximately 4220°F. An oxyacetylene torch is generally used for making joints in copper pipes 1/2 in or larger in diameter. See Table M-2-1.

FUNDAMENTALS OF HVAC/R TEXT REFERENCE
Unit 16

Required Tools and Equipment

Air acetylene torch	Unit 16
Copper piping, fitting, brazing rod	Units 15 and 16
Hacksaw	Unit 9

SAFETY REQUIREMENTS

A. One hundred percent cotton or leather clothing is the best material to wear while brazing, soldering, or welding.

B. Shirts should have long sleeves and work gloves must be worn (cloth, leather palm, or all leather).

C. When using acetylene, the torch pressure should be approximately 5 psig and the cylinder valve should be open no more than 1 1/2 turns.

TABLE M-2-1 Common soldering and brazing metal and fluxes showing base metals that can be joined

	Alloy	Flux type	Base metal
S O L D E R	95-5 Tin antimony solder[1]	C-Flux	Copper pipe, brass, steel
	95-5 Tin antimony solder	Rosin	Copper pipe, copper wiring, brass
	95-5 Tin antimony solder	Acid	Copper pipe, brass, steel, galvanized sheet metal
	98-2 Tin silver solder	Mineral based flux	Copper pipe, brass, steel
	40-60 Cadmium zinc solder	Specific flux from solder manufacturer	Aluminum
B R A Z I N G	Copper phosphorus silver brazing BCuP	1% to 15% Silver no flux required	Copper pipe
	Copper phosphorus silver brazing BCuP	1% to 15% Silver mineral based flux	Copper pipe to brass, brass, steel
	Copper silver brazing BCuP	45% Silver mineral flux	Copper pipe to steel, brass, steel

[1]The percentages of the materials in the flux are given in the numbers, for example 95% tin, 5% antimony.

PROCEDURE

STEP 1. Prepare the copper piping and fitting for use.

   A. Use an abrasive sanding cloth or wire brush to remove all contaminants from the surface of the pipe. See Figure M-2-1 and Figure M-2-2.

   B. Do not touch the cleaned surfaces with your hands as oil from your fingers can prevent solder from flowing completely into a joint.

   C. Use a brush to apply the flux to the end of the pipe.

   D. *Do not apply flux to the very end of the pipe.*

   E. Apply the flux to approximately 1/16 in to 1/8 in from the end of the pipe to avoid flux contamination into the system.

STEP 2. Prepare to light the torch as follows:

   A. Once the air acetylene torch and regulator have been attached to the acetylene cylinder, the acetylene cylinder valve is opened one quarter turn.

   B. OSHA requires that a nonadjustable wrench be used on any cylinder valve stems that are not equipped with a hand wheel.

   C. Hold the cylinder steady with one hand while opening the valve with the other hand.

   D. Never open the acetylene cylinder valve more than 1 1/2 turns.

   E. With the pressure adjusted properly according to the manufacturer's recommendations for that torch tip, turn the acetylene valve slightly to light the flame.

   F. The only safe devices to use to light any torch are those specifically designed for that purpose, such as spark lighters or flint lighters.

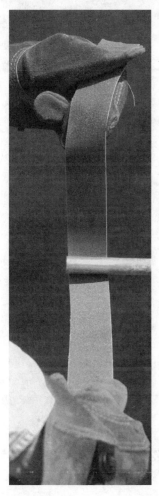

Figure M-2-1

Figure M-2-2

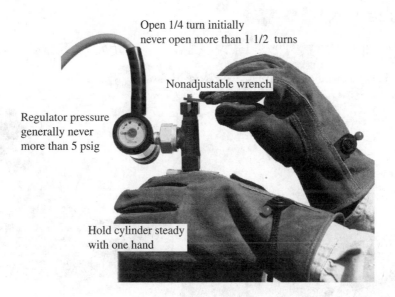

Open 1/4 turn initially
never open more than 1 1/2 turns

Nonadjustable wrench

Regulator pressure
generally never
more than 5 psig

Hold cylinder steady
with one hand

Figure M-2-3

G. When using a flint lighter hold it slightly off to one side of the torch tip. See Figure M-2-4. Holding the lighter over the end of the tip may cause a pop when the torch is lit.

H. Open the hand wheel of the torch body completely.

Figure M-2-4

I. Failure to provide the torch tip with the proper gas flow will result in the tip being overheated.

J. With low gas flows rates, the flame gets closer and closer to the tip and it will become hot.

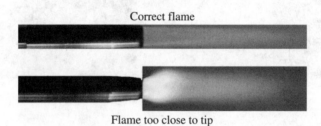

Figure M-2-5

STEP 3. Begin soldering the joint as follows:

A. Heat the pipe first and then the fitting. See Figure M-2-6.

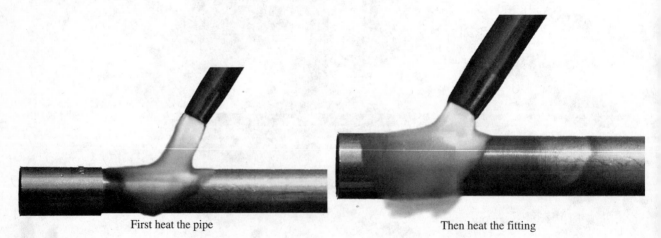

Figure M-2-6

B. Once the pipe has become hot, approaching a dull red color, move the torch flame onto the fitting so that it envelopes both the fitting and the pipe.

C. Occasionally touch the pipe surface with the tip of the brazing rod as a test of temperature readiness.

D. Continue heating the pipe until the brazing metal begins to flow evenly over the surface. See Figure M-2-7.

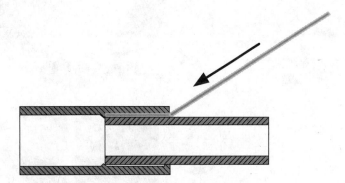

Figure M-2-7

E. If you watch carefully, you can see a slight change in the fitting's color as the filler metal flows into the joint space.

F. Bringing the torch down on the fitting will help draw the filler metal completely into the joint.

G. Once the joint gap has been filled, continue adding small amounts of filler braze metal until a fillet of metal surrounds the joint.

STEP 4. Destructively test the joint as follows:

A. Use a hacksaw to saw off the copper fitting at a point just beyond the depth that the pipe was inserted into the joint.

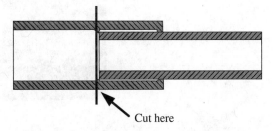

Cut here

Figure M-2-8

B. Once the fitting has been cut completely apart, clamp the pipe in a vise and cut straight down through the entire joint into the pipe.

C. Rotate the pipe 90° and repeat the process so that there are four cuts. See Figure M-2-9.

D. Bend each quarter out with a wrench.

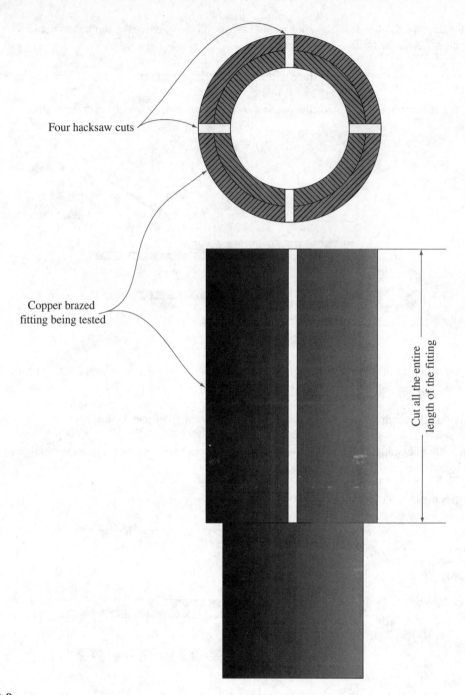

Four hacksaw cuts

Copper brazed
fitting being tested

Cut all the entire
length of the fitting

Figure M-2-9

Figure M-2-10

E. Using a hammer and an anvil, flatten each of the four corners. See Figure M-2-10.

F. As the joint pieces are flattened, it is easy to see areas where 100% joint penetration did not occur.

QUESTIONS

To get help in answering some of the following questions, refer to the *Fundamentals of HVAC/R* text Unit 16

(Circle the letter that indicates the correct answer.)

1. Acetylene gas must be stabilized in the cylinder with:
A. propane.
B. oxygen.
C. acetone.
D. nitrogen.

2. It is illegal to operate a torch with an acetylene pressure:
   A. greater than 15 psi.
   B. less than 15 psi.
   C. greater than 5 psi.
   D. greater than 3 psi.

3. Turning the acetylene pressure up beyond the recommended pressure value for a specific torch will:
   A. make the torch operate better.
   B. cause it to "pop."
   C. not make it work better.
   D. cause it to "crack."

4. It is against OSHA regulations to open any acetylene cylinder more than:
   A. 1 1/2 turns.
   B. 1 turn.
   C. 2 turns.
   D. 2 1/2 turns.

5. When transporting an acetylene cylinder:
   A. lay it on its side.
   B. leave the cylinder "cracked" open.
   C. always remove the regulator.
   D. leave the cylinder cap off.

BELT DRIVE BLOWER COMPLETE SERVICE

LABORATORY OBJECTIVE

The purpose of this laboratory worksheet is to demonstrate your ability to conduct the proper maintenance service procedure for a belt driven blower.

LABORATORY NOTES

Some older residential and most commercial air handlers use belt drive blowers to deliver air through the duct system. The advantage of belt drive blowers over direct drive is the flexibility in choosing blower speed and the ease of manufacturing a blower wheel to withstand the rpm. It is too difficult to match all the factors, rpm, CFM, motor hp, and so on, in all commercial applications. Most new residential systems have gone over to direct drive blowers but in commercial systems, belt drives will be around for many years to come. Occasionally a belt drive blower needs a complete overhaul including any or all of the following: motor and bearings, motor pulley, belt, blower pulley, blower bearings, blower shaft, cleaning, balancing, or rebalancing.

FUNDAMENTALS OF HVAC/R TEXT REFERENCE
Unit 69

Required Tools and Equipment

Belt driven blower assembly	
Tool kit	Unit 10
Multimeters	Unit 11
Clamp-on ammeter	Unit 11

SAFETY REQUIREMENTS

A. Check all circuits for voltage before doing any service work.

B. Stand on dry nonconductive surfaces when working on live circuits.

C. Never bypass any electrical protective devices.

PROCEDURE

STEP 1. Collect the blower data and fill in the chart.

Blower Unit Data

Unit name:			Model number:	Blower motor amperage rating:	
Unit type:					
Blower motor data:	Voltage:		Horsepower:	Locked rotor amps:	Rated load amps:
					Full load amps:
Blower wheel data:	Width:		Diameter:	Shaft size OD:	
Pulley size:	Motor pulley OD:		Motor pulley ID:	Blower shaft OD:	Blower shaft ID:
Belt size:			**Belt number:**		
Lubrication required for blower bearings: **(circle one)**				Grease	Oil
Lubrication required for motor bearings: (circle one)				Grease	Oil
Inspect greased motors for grease relief fitting at bottom of bearing below grease opening:				None _____	

STEP 2. Complete an initial inspection. Complete the Initial Inspection Checklist and make sure to check each step off in the appropriate box as you finish it. This will help you to keep track of your progress.

Initial Inspection Checklist

Step	Procedure	Check
1	Spin the blower wheel slowly by hand.	
2	Notice any drag noise or pulley wobble.	
3	Install the clamp-on ammeter to the L1 terminal of the motor.	
4	Place the blower door in the normal position for normal airflow and load on the motor.	
5	Obtain the fan only operation and record the motor amps. Amps = _____	
6	Remove blower door for further inspection.	
7	Observe the motor begin to turn and watch for belt slipping during startup.	
8	Note any running noise, bearing, grinding, or any noise other than normal airflow noise.	

STEP 3. After finishing the initial blower inspection, make sure all power is off. Lock and tag the power panel before removing any parts. Complete the Blower Removal and Cleaning Checklist.

Blower Removal and Cleaning Checklist

Step	Procedure	Check
1	Make sure that the power to the blower motor is secured. Verify this with a multimeter voltage test. Refer to Laboratory Worksheet E-1.	
2	Remove and carefully store all panels and covers.	
3	Loosen tension on belt for removal.	
4	Remove belt and inspect for any cracks or severe shine (caused by slipping).	
5	Inspect pulleys for wear and grooving.	
6	Inspect blower wheel blades and scrape with a screwdriver or thin tool for any evidence of accumulated debris.	
7	Clean blower wheel with water, compressed air, or CO_2, whichever is most available and appropriate.	
8	Blow motor air passages with compressed air or CO_2.	
9	Wipe down the motor and name plate with a clean rag.	

STEP 4. After cleaning, the blower assembly can be disassembled. Complete the Blower Disassembly Checklist.

Blower Disassembly Checklist

Step	Procedure	Check
1	Loosen setscrew from the blower pulley and remove the pulley.	
2	Support the blower wheels to keep them in position.	
3	Loosen the screws on the set collars and pull the blower shaft out of the blower.	
4	Record the length and diameter of the blower shaft. Length = _____ OD = _____	
5	List any special features of blower keyways, flat sides, and so on, for replacement.	
6	Make notes on the blower shaft condition such as wear, grooving, rust, and so on.	
7	Roll the shaft on a flat surface to be sure it is straight and true.	
8	Remove and inspect blower wheel bearings and replace if necessary.	
9	Install the original or a replacement blower shaft.	

(Continued)

Blower Disassembly Checklist   (*Continued*)

Step	Procedure	Check
10	Install the original or replacement shaft pulley.	
11	Install the original or replacement shaft pulley.	
12	Align or check alignment of pulleys with a straight edge on the pulley outer edge. The straight edge should touch an outer edge of both pulleys in at least four different places.	
13	Lubricate the motor and shaft bearings.	
14	Install a new belt of the correct size.	
15	Adjust the motor tension assembly for the correct belt tension. Consult the manufacturer's recommendation for correct belt tension (belt play is typically 2–3 in).	

STEP 5. After reassembling the blower, you can check for correct operation with a final test run. Complete the Blower Test Run Checklist.

Blower Test Run Checklist

Step	Procedure	Check
1	Spin the blower wheel slowly by hand.	
2	Notice any drag noise or pulley wobble.	
3	Install the clamp-on ammeter to the L1 terminal of the motor.	
4	Turn on the main power.	
5	Place the blower door in the normal position for normal airflow and load on the motor.	
6	Obtain the fan only operation and record the motor amps. Amps = _____	
7	Compare measured amps with rated motor amps. Actual amps is: (circle one) Higher Lower	
8	If the running amperage is incorrect then loosen the motor belt tension and remove the belt.	
9	Loosen and adjust the motor pulley or replace with a pulley of larger or smaller size as required. Move the pulley sheave closer when installing a larger pulley.	
10	Install the belt and adjust the belt tension.	
11	Run the blower and once again measure the amperage. Amps = _____	
12	Repeat steps 8 through 11 to obtain rated full load amps and maximum CFM.	

WATER CIRCULATING PUMP COMPLETE SERVICE

LABORATORY OBJECTIVE

The purpose of this laboratory worksheet is to demonstrate your ability to conduct the proper maintenance service procedure for a water circulating pump.

LABORATORY NOTES

Hot water heating boilers are dependent on water pumps to distribute the hot water they produce. The pumps are primarily driven by electric motors. The electric motor requires typical motor service, cleaning, lubrication, proper mounting, insulation resistance, and amperage checks. The motor is usually connected to the pump with a spider type coupler or flexible spring connected coupler. Improper alignment is the main reason for coupler problems. Other pump maintenance issues fall into two major categories, which are impeller or shaft seal problems.

On smaller pumps it is typical to exchange the entire pump or bearing assembly for a new or rebuilt replacement. Larger pumps can generally be brought in to be rebuilt by the local factory representative. If the pump is to be rebuilt on site, it is recommended that you have attended factory service demonstrations and have been trained in this area.

FUNDAMENTALS OF HVAC/R TEXT REFERENCE
Unit 71

Required Tools and Equipment

Circulating pump	Unit 71
Tool kit	Unit 10
T-handle Allen wrench	Unit 10
Strap wrench	Unit 10
Compressed air	
Cleaner-degreaser	
Rubber hammer	Unit 10
Thin chisel	Unit 10

SAFETY REQUIREMENTS

A. Check all circuits for voltage before doing any service work.

B. Be careful to avoid lubricant contact with exposed skin.

C. Always wear safety glasses whenever using compressed air or CO_2 for blowing out dust and dirt.

PROCEDURE

STEP 1. Collect the pump data and fill in the chart.

Blower Unit Data

Circulating pump make:	Model number:	
Motor data:	Horsepower:	Type and size of coupler:

STEP 2. Complete an initial inspection. Complete the Initial Inspection and Cleaning Checklist and make sure to check each step off in the appropriate box as you finish it. This will help you to keep track of your progress.

Initial Inspection and Cleaning Checklist

Step	Procedure	Check
1	Make sure that the power to the circulating pump motor is secured. Verify this with a multimeter voltage test. Refer to Laboratory Worksheet E-1.	
2	Use CO_2 or compressed air pressure to blow out and clean all open air passages of the motor and bearing assembly.	
3	Use a nondetergent or recommended lubricant for the motor and bearing assembly.	
4	Using a pump sprayer, apply a cleaner-degreaser to the motor and pump assembly.	
5	Use cleaning rags to wipe off all excess oil and any accumulated dirt, dust, grease, and so on.	
6	Inspect the bearing assembly for any water dripping, metal shavings, scraping or grinding noise, and so on.	
7	Inspect motor mounts. Is the motor centered within the rubber mount or is the motor sagging in the rubber?	
8	If the motor is sagging or out of center in the rubber mount, the motor mounts must be replaced.	

STEP 3. After finishing the initial circulating pump inspection and cleaning, make sure all power is off. Lock and tag the power panel before removing any parts. Complete the Motor Mount Replacement Checklist.

Motor Mount Replacement Checklist

Step	Procedure	Check
1	Make sure that the power to the blower motor is secured. Verify this with a multimeter voltage test. Refer to Laboratory Worksheet E-1.	
2	Using a thin open end wrench, loosen the two top motor mount bolts from inside the bearing assembly.	
3	Loosen and remove the two bottom motor mount bolts.	
4	Use a long T-handle Allen wrench to remove the motor end of the spring coupler.	
5	Supporting the motor in one hand, remove the previously loosened top two motor mount bolts.	
6	Remove the motor from its cradle and the two machine screws connected to the motor mount straps.	
7	Using an old screwdriver or a thin chisel, pry off both rubber motor mounts from the motor.	
8	Use a small metal rubber or plastic hammer to tap in place two new rubber motor mounts.	

STEP 4. If the bearings drag and need to be replaced, then complete the Bearing Replacement Checklist.

Bearing Replacement Checklist

Step	Procedure	Check
1	Make sure that the power to the blower motor is secured. Verify this with a multimeter voltage test. Refer to Laboratory Worksheet E-1.	
2	Turn off the water isolation valves for the pump and drain off any remaining water.	
3	Remove the motor as in Step 3 from the Motor Mount Replacement Checklist.	
4	Remove the bolts holding the bearing assembly to the pump housing.	
5	Remove the bearing assembly along with the pump impeller.	
6	Inspect and clear any debris from the pump housing.	
7	Hold the impeller with a strap wrench and use a socket wrench to remove the bolt holding the impeller to the pump shaft.	
8	Inspect the impeller and replace it if there are any signs of wear or deterioration.	
9	Install the impeller on the new bearing assembly.	
10	Remove any old gaskets and carefully scrape and clean the surface metal. Always install new gaskets.	

Bearing Replacement Checklist (*continued*)

Step	Procedure	Check
11	After installing new gasket, slip the bearing assembly into the pump housing.	
12	Install and tighten the bolts holding the bearing assembly to the pump housing.	
13	Install the coupler to the bearing assembly shaft.	
14	Hold the motor in position and slide the coupler onto the motor shaft. Be sure to tighten the setscrews into the hollow spot on both the pump and motor shafts.	
15	While still holding the motor, start the two top motor mount bolts first, then the bottom two bolts.	
16	Tighten the motor mount bolts.	
17	Open the water isolation valves and obtain normal pressure on the system. Purge any air from the system.	
18	Start up the pump and obtain normal operation.	
19	Lubricate the bearing assembly and motor as required.	
20	Inspect the bearing assembly for any water leaks.	

GAS FURNACE STARTUP

LABORATORY OBJECTIVE

The purpose of this Laboratory Worksheet is to demonstrate your ability to go through the necessary sequence for a typical gas furnace startup.

LABORATORY NOTES

You will need to familiarize yourself with gas furnace arrangements and types. Always remember to make the visual check of all system components prior to starting a gas furnace.

FUNDAMENTALS OF HVAC/R TEXT REFERENCE
Unit 40

Required Tools and Equipment

Operating gas furnace	Unit 38
Tool kit	Unit 10
Multimeters	Unit 11
Clamp-on ammeter	Unit 11

SAFETY REQUIREMENTS

A. Never allow gas flow without a flame. If gas is allowed to build up and then suddenly ignite, this can create a serious hazard.

B. If the furnace contains a pilot, always ensure it is properly lit prior to starting the unit.

C. Check all circuits for voltage before doing any service work.

D. Stand on dry nonconductive surfaces when working on live circuits.

E. Never bypass any electrical protective devices.

PROCEDURE

STEP 1. Collect the gas furnace data and fill in the chart.

Gas Furnace Unit Data

Furnace make:		Model number:	Blower motor amperage rating:	
Furnace type (circle one)	Upflow	Counterflow	Basement	Horizontal
Blower type (circle one)	Direct drive	Belt drive		
Blower speed (circle one)	Single speed	Two speed	Three speed	Four speed
System type (circle one)	Heat only	Heat and humidify	Heat and cool	
Burner type (circle one)	Atmospheric	Induced draft	Power	

STEP 2. Complete the Prestart Checklist and make sure to check each step off in the appropriate box as you finish it. This will help you to keep track of your progress.

Prestart Checklist

Step	Procedure	Check
1	Turn the thermostat down below room temperature.	
2	Make sure all power to the furnace is off. Lock and tag the power panel before removing any parts.	
3	Vent connector connected with three screws per joint.	
4	Fuel line installed properly, with no apparent leaks.	
5	Spin all fans to be sure they are loose and turn freely.	
6	Combustion area free from debris.	
7	Electrical connections complete—main power.	
8	Electrical connections complete—thermostat.	
9	All doors and panels available and in place.	
10	Thermostat installed and operating correctly.	

STEP 3. After finishing the Prestart Checklist you are now ready to start the furnace and check its operation. You will follow the same procedure of working through a checklist. This will help you to keep track of your progress. Complete the Startup Checklist and make sure to check each step off in the appropriate box as you finish it.

Startup Checklist

Step	Procedure	Check
1	Check the main fuse or breaker and the amperage rating and make sure it is the correct rating for the furnace.	
2	Measure the incoming supply voltage. Refer to Laboratory Worksheet E-1 for the proper procedure for testing the circuit.	
3	If the Steps 1 and 2 have been completed and are satisfactory, then turn on the supply power for the furnace.	
4	Set the thermostat to FAN ON to obtain the fan only operation and observe that the blower is operating. If the thermostat is not so equipped then consult with your instructor.	
5	Light the pilot if so equipped, following the manufacturer's instructions provided with the furnace.	
6	Turn the fan off and the thermostat to heat.	
7	Adjust the thermostat setting to 10°F above the room temperature.	
8	Observe as the flame sequence begins and the flame comes on.	
9	Observe that the blower begins running after the heat exchanger comes up to temperature.	
10	Use a clamp-on ammeter to measure the blower motor amperage and compare this value to its rating. Refer to Laboratory Worksheet E-1 for the proper procedure for using electrical meters.	
11	Turn the thermostat setting down below the room temperature and observe the flame go off.	
12	Observe the blower stop running approximately 3 min after the burner turns off.	

MEASURE GAS USAGE

LABORATORY OBJECTIVE

The purpose of this lab is to demonstrate your ability to determine the gas usage for a furnace.

LABORATORY NOTES

You will need to familiarize yourself with gas furnace arrangements and types. Always remember to make a visual check of all system components prior to starting a gas furnace.

FUNDAMENTALS OF HVAC/R TEXT REFERENCE

Unit 40

Required Tools and Equipment

Operating gas furnace	Unit 38
Tool kit	Unit 10
Multimeters	Unit 11
Clamp-on ammeter	Unit 11

SAFETY REQUIREMENTS

A. Never allow gas flow without a flame. If gas is allowed to build up and then suddenly ignite, this can create a serious hazard.

B. If the furnace contains a pilot, always ensure it is properly lit prior to starting the unit.

C. Check all circuits for voltage before doing any service work.

D. Stand on dry nonconductive surfaces when working on live circuits.

E. Never bypass any electrical protective devices.

PROCEDURE

STEP 1. Collect the gas furnace data and fill in the chart.

Gas Furnace and Fuel Meter Data

Furnace make:		Model number:	Blower motor amperage rating:	
Burner make:				
Number of main burner orifices:				
Input rating in Btu/hr:				
Burner type (circle one)	Atmospheric	Induced draft	Power	
Meter type (circle one)	½ ft³	2 ft³		

STEP 2. Complete the Prestart Checklist and prepare to start the furnace as outlined in the procedure provided in Laboratory Worksheet GH-1.

STEP 3. After finishing the Prestart Checklist and then starting the furnace to check its operation, you can prepare to clock the main burner. You will follow the same procedure of working through a checklist. This will help you to keep track of your progress. Complete the Clock Main Burner Checklist and make sure to check each step off in the appropriate box as you finish it.

Clock Main Burner Checklist

Step	Procedure	Check
1	Isolate the main burner.	
2	Leave the pilot on for normal operation.	
3	Operate the main burner to be tested.	
4	Observe the rotating dial of the gas meter. This is typically either a ½ ft³ or 2 ft³ dial	
5	Measure the time in seconds for one revolution of the dial. Time in seconds =	
6	Calculate cubic feet per hour (CFH). CFH = 3,600/seconds × ½ ft³ or 2 ft³ dial =	
7	Calculate actual burner input. The approximate heating value for natural gas is 1,000 Btu per ft³. You may contact the gas supplier to obtain a more accurate value if desired. Actual burner input = CFH × 1,000 Btu/ft³ = BTUH	
8	Compare actual measured input with the name plate input rating.	
9	If there is a difference of greater than 5% between the actual measured input as compared to the name plate input rating, then the gas manifold pressure should be checked.	

GAS FURNACE PREVENTIVE MAINTENANCE (PM)

LABORATORY OBJECTIVE

The purpose of this laboratory worksheet is to demonstrate your ability to conduct the proper preventive maintenance procedure for a gas furnace.

LABORATORY NOTES

You will need to familiarize yourself with gas furnace arrangements and types. Always remember to make the visual check of all system components prior to starting a gas furnace.

FUNDAMENTALS OF HVAC/R TEXT REFERENCE
Unit 40

Required Tools and Equipment

Operating gas furnace	Unit 38
Tool kit	Unit 10
Multimeters	Unit 11
Clamp-on ammeter	Unit 11

SAFETY REQUIREMENTS

A. Never allow gas flow without a flame. If gas is allowed to build up and then suddenly ignite, this can create a serious hazard.

B. If the furnace contains a pilot, always ensure it is properly lit prior to starting the unit.

C. Check all circuits for voltage before doing any service work.

D. Stand on dry nonconductive surfaces when working on live circuits.

E. Never bypass any electrical protective devices.

PROCEDURE

STEP 1. Collect the gas furnace data and fill in the chart.

Gas Furnace Unit Data

Furnace make:		Model number:	Blower motor amperage rating:	
Furnace type (circle one)	Upflow	Counterflow	Basement	Horizontal
Blower type (circle one)	Direct drive	Belt drive		
Blower speed (circle one)	Single speed	Two speed	Three speed	Four speed
System type (circle one)	Heat only	Heat and humidify	Heat and cool	
Burner type (circle one)	Atmospheric	Induced draft	Power	

STEP 2. Complete the Prestart Checklist and prepare to start the furnace as outlined in the procedure provided in Laboratory Worksheet GH-1. With the furnace operating complete the Running Checklist.

Running Checklist

Step	Procedure	Check
1	Observe a normal sequence of operation (burner on, fan on, burner off, fan off). Check for any unusual noise such as bearing noise.	
2	Inspect operating pilot and note the color, size, shape, and position.	
3	Inspect burner light off for flame lifting, floating, noise, and smooth ignition.	
4	Measure gas manifold pressure.	

STEP 3. After finishing the Running Checklist, shut down the furnace. Make sure all power to the furnace is off. Lock and tag the power panel before removing any parts. Complete the Blower Maintenance Checklist.

Blower Maintenance Checklist

Step	Procedure	Check
1	Make sure that the power to the blower motor is secured. Verify this with a multimeter voltage test. Refer to Laboratory Worksheet E-1.	
2	Remove the wires from the blower motor at an accessible location.	

Step	Procedure	Check
3	Remove screws or bolts securing the blower assembly.	
4	Inspect belt for cracks and signs of wear.	
5	Inspect pulleys for wear, grooving, and alignment.	
6	Spin blower by hand and observe pulleys turn.	
7	Inspect for pulley wobble and alignment.	
8	Listen for bearing noise, drag, or movement.	
9	Inspect blower shaft for signs of wear.	
10	Clean blower and motor with air pressure, brushes, scrapers, and cleaning solution as required.	
11	Reassemble blower, taking care to check pulley alignment and correct belt tension.	
12	Check the condition of the heat exchanger before reinstalling the blower.	

STEP 4. After finishing the Blower Maintenance Checklist you can perform the necessary burner maintenance. Again, make sure all power to the furnace is off. Lock and tag the power panel before removing any parts. Complete the Gas Burner Maintenance Checklist.

Gas Burner Maintenance Checklist

Step	Procedure	Check
1	Remove and clean pilot assembly.	
2	Clean and inspect flame sensor and igniter.	
3	Remove and clean main burners, inspect and mark burners for original location. Note: All burners are not interchangeable.	
4	Remove vent connector, draft diverter, and any flue baffles.	
5	Vacuum and brush all soot, rust, and solid particles from the fire side of the heat exchanger.	
6	Insert a light into the combustion area.	
7	If possible turn off the lights in the furnace room.	
8	With the light in each burner, inspect the heat exchanger from both the fan side and the plenum side for holes.	
9	Reinstall the blower assembly after the heat exchanger check.	
10	Reinstall main and pilot burners.	

Step	Procedure	Check
11	Reconnect vent components.	
12	Prepare to start the furnace for a running check following the startup procedure from Laboratory Worksheet GH-1.	
13	Light pilot and adjust for proper size, configuration, and location.	
14	Turn on main burner and adjust gas pressure as required.	
15	Adjust air shutter to obtain correct flame color and CO_2.	

GAS FURNACE COMBUSTION TESTING

LABORATORY OBJECTIVE

The purpose of this lab is to demonstrate your ability to test a gas furnace to determine proper combustion.

LABORATORY NOTES

Combustion testing is primarily used when setting power burners on which there is an adjustment that will allow total overcombustion to the burners. The traditional standard setting is at 50% excess air to ensure complete combustion. Consult the manufacturer's recommendation for the burner, furnace, or boiler type for the more accurate settings. When using one of the newer electronic combustion analyzers, the appearance of CO_2 will show when complete combustion is taking place.

FUNDAMENTALS OF HVAC/R TEXT REFERENCE

Units 40 and 13

Required Tools and Equipment

Operating gas furnace	Unit 38
Tool kit	Unit 10
Combustion analyzer	Unit 13
Temperature sensor	Unit 12

SAFETY REQUIREMENTS

A. Never allow gas flow without a flame. If gas is allowed to build up and then suddenly ignite, this can create a serious hazard.

B. If the furnace contains a pilot, always ensure it is properly lit prior to starting the unit.

C. Vent pipes will be hot with the furnace operating. Be careful not to touch hot surfaces when inserting, removing, or reading instruments.

PROCEDURE

Step 1. Collect the gas furnace data and fill in the chart.

Gas Furnace Unit Data

Furnace make:		Model number:		Blower motor amperage rating:	
Furnace type (circle one)	Upflow	Counterflow	Basement	Horizontal	
Blower type (circle one)	Direct drive	Belt drive			
Blower speed (circle one)	Single speed	Two speed	Three speed	Four speed	
System type (circle one)	Heat only	Heat and humidify	Heat and cool		
Burner type (circle one)	Atmospheric	Induced draft	Power		

Step 2. Complete the Prestart Checklist and prepare to start the furnace as outlined in the procedure provided in Laboratory Worksheet GH-1. Before starting the furnace, complete the Preparation for Combustion Test Checklist.

Preparation for Combustion Test Checklist

Step	Procedure	Check
1	Prior to starting the furnace, locate CO_2 test openings and the temperature probe location for an undiluted vent gas sample.	
2	Insert the thermometer probe and check the CO_2 for position.	
3	Start the furnace and allow it to run until the vent gas temperature is at its normal maximum.	

Step 3. With the furnace now operating, begin taking your initial readings and record them in the Combustion Test Chart.

  A. Initial readings are taken prior to making any adjustments.

  B. For the yellow flame test, close the primary air shutter until a lazy flame is present.

  C. For the excess air test, open the primary air shutter to a maximum amount.

  D. For the 8% CO_2 test, close the air shutter to obtain 8%.

  E. Calculate the combustion efficiency using a slide rule calculator with the natural gas slide (see *Fundamentals of HVAC/R* text, Unit 40).

Combustion Test Chart

Combustion Test Readings	Test 1 (initial)	Test 2 (yellow)	Test 3 (excess)	Test 4 (8% CO_2)
CO_2				
Actual stack temperature				
Net stack (gross–room)				
Combustion efficiency				

F. Which flame is the most efficient?

_____

_____

Name _____

Date _____

Instructor's OK ☐

MEASURE GAS FURNACE THERMAL EFFICIENCY

LABORATORY OBJECTIVE

The purpose of this lab is to demonstrate your ability to measure the thermal efficiency of a gas furnace and to draw a correlation, if there is any, between the airflow through the furnace and the thermal efficiency of the furnace.

LABORATORY NOTES

Thermal efficiency is Btu output divided by Btu input. Input is the measure of gas consumed while output is measured by the sensible heat formula, which is (temperature in °F) × (1.08) × (airflow in CFM). The thermal efficiency will be calculated at three different airflows.

FUNDAMENTALS OF HVAC/R TEXT REFERENCE
Unit 40

Required Tools and Equipment

Operating gas furnace with variable speed blower	Unit 38
Tool kit	Unit 10
Temperature sensor	Unit 12
Airflow hood	

SAFETY REQUIREMENTS

A. Never allow gas flow without a flame. If gas is allowed to build up and then suddenly ignite, this can create a serious hazard.

B. If the furnace contains a pilot, always ensure it is properly lit prior to starting the unit.

C. Vent pipes will be hot with the furnace operating. Be careful not to touch hot surfaces when inserting, removing, or reading instruments.

PROCEDURE

STEP 1. Collect the gas furnace data and fill in the chart.

Gas Furnace Unit Data

Furnace make:			Model number:	Blower motor amperage rating:	
Furnace type (circle one)	Upflow	Counterflow	Basement	Horizontal	
Blower type (circle one)	Direct drive	Belt drive			
Blower speed (circle one)	Single speed	Two speed	Three speed	Four speed	
System type (circle one)	Heat only	Heat and humidify	Heat and cool		
Burner type (circle one)	Atmospheric	Induced draft	Power		

STEP 2. Complete the Prestart Checklist and prepare to start the furnace as outlined in the procedure provided in Laboratory Worksheet GH-1. Before starting the furnace complete the Preparation for Efficiency Test Checklist.

Preparation for Efficiency Test Checklist

Step	Procedure	Check
1	Prepare to measure the airflow on the inlet and/or the outlet of the furnace using an airflow hood. It may be necessary to build a bracket to install the airflow hood.	
2	If the airflow hood is located on the outlet, the airflow measurement must be taken prior to starting the furnace as the airflow hood will be damaged if exposed to temperatures above 140°F (fan only operation).	
3	Operate the furnace in the normal heating mode. The airflow hood may be left on the return air during this test. Airflow will typically go down as the air is heated due to the air restriction of the furnace and the increased air volume of the heated air.	

STEP 3. With the furnace now operating, begin taking your initial readings and record them in the Efficiency Test Chart.

    A. Initial readings are taken prior to making any adjustments.

    B. For the reduced airflow test, reduce the blower speed or throttle the air damper.

    C. For the increased airflow test, increase the blower speed or open the air damper.

    D. For the 8% CO_2 test, close the air shutter to obtain 8%.

    E. Calculate the combustion efficiency using a slide rule calculator with a natural gas slide (see *Fundamentals of HVAC/R* text, Unit 40).

Efficiency Test Chart

Efficiency test readings	Test 1 (initial)	Test 2 (reduced airflow)	Test 3 (increased airflow)
Air inlet temperature			
Air outlet temperature			
Airflow (CFM)			

STEP 4. Calculate the Btu output for the three test conditions using the formula:
Btu output = (outlet temperature − inlet temperature) × (1.08) × (airflow in CFM)

Btu Output

	Test 1 (initial)	Test 2 (reduced airflow)	Test 3 (increased airflow)
Btu output			

STEP 5. Calculate the efficiency for the three test conditions using the formula:
Efficiency = (Btu output from Step 4)/(burner rated Btu input)

Calculated Efficiency

	Test 1 (initial)	Test 2 (reduced airflow)	Test 3 (increased airflow)
Calculated efficiency			

A. What correlation can be drawn between the thermal efficiency and the airflow of the furnace?

_____

_____

_____

PILOT TURNDOWN TEST—THERMOCOUPLE TYPE

LABORATORY OBJECTIVE
The purpose of this lab is to test the proper operation of a thermocouple type pilot for a gas furnace.

LABORATORY NOTES
A pilot turndown test determines the smallest possible pilot capable of proving flame presence to the control circuit and lighting the main burner safely. It is performed by lighting the pilot, measuring the pilot flame signal, and observing a series of main burner ignitions using various pilot sizes. Remember that a cold burner is more difficult to ignite and in the performance of this test the burner becomes warmed up; you must allow sufficient cooldown time to perform the final test. A pilot turndown test can also point to other pilot problems.

FUNDAMENTALS OF HVAC/R TEXT REFERENCE
Unit 39

Required Tools and Equipment

Operating gas furnace	Unit 38
Tool kit	Unit 10
Millivoltmeter	Unit 11

SAFETY REQUIREMENTS

A. Never allow gas flow without a flame. If gas is allowed to build up and then suddenly ignite, this can create a serious hazard.

B. If the furnace contains a pilot, always ensure it is properly lit prior to starting the unit.

C. Check all circuits for voltage before doing any service work.

PROCEDURE

STEP 1. Obtain any standard thermocouple installed or not installed. Complete the Thermocouple Open Circuit Test (30 MV) Checklist and make sure to check each step off in the appropriate box as you finish it. This will help you to keep track of your progress.

Thermocouple Open Circuit Test (30 MV) Checklist

Step	Procedure	Check
1	Remove the threaded connection from the gas valve or pilot safety switch if necessary.	
2	Connect a millivoltmeter from the outer copper line to the inner core lead at the end of the thermocouple.	
3	Heat the enclosed end of the thermocouple with a torch or a normal pilot assembly if installed.	
4	A voltage of up to 30 mV will be read on the meter. If the meter goes down, then reverse the leads.	
5	Record the highest millivoltage. Voltage = _____ mV	

STEP 2. Obtain a thermocouple installed on a standard combination gas valve or pilot safety switch. Complete the Thermocouple Closed Circuit Test (30 MV) Checklist.

Thermocouple Closed Circuit Test (30 MV) Checklist

Step	Procedure	Check
1	Obtain the thermocouple adaptor for the valve end.	
2	Light the pilot for a normal pilot flame.	
3	Read the millivolts at the adaptor. Voltage = _____ mV	
4	Blow out the pilot light and then relight it within 30 s of going out.	
5	Why does gas still come out of the pilot burner after the pilot is out?	
6	Blow out the pilot again and observe the voltage output.	
7	Record the voltage in millivolts when the pilot valve closes stopping the pilot gas so that the pilot will not relight. This value should be between 9 and 12 mV. Voltage = _____ mV	
8	Adjust the size of the pilot flame for the smallest pilot flame capable of proving the pilot and lighting the main burner smoothly and safely.	

PILOT TURNDOWN TEST—FLAME ROD TYPE

LABORATORY OBJECTIVE

The purpose of this lab is to test the proper operation of a flame rod type pilot for a gas furnace. Perform this test on any furnace equipped with a pilot, a separate spark igniter, and a flame rod.

LABORATORY NOTES

A pilot turndown test is a test to determine the smallest possible pilot capable of proving flame presence to the control circuit and lighting the main burner safely. It is performed by lighting the pilot, measuring the pilot flame signal, and observing a series of main burner ignitions using various pilot sizes. Remember that a cold burner is more difficult to ignite and in the performance of this test the burner becomes warmed up; you must allow sufficient cooldown time to perform the final test. A pilot turndown test can also point to other pilot problems.

FUNDAMENTALS OF HVAC/R TEXT REFERENCE

Unit 39

Required Tools and Equipment

Operating gas furnace	Unit 38
Tool kit	Unit 10
Micro-ammeter	Unit 11

SAFETY REQUIREMENTS

A. Never allow gas flow without a flame. If gas is allowed to build up and then suddenly ignite, this can create a serious hazard.

B. If the furnace contains a pilot, always ensure it is properly lit prior to starting the unit.

C. Check all circuits for voltage before doing any service work.

PROCEDURE

STEP 1. Perform this test on any furnace equipped with a pilot, a separate spark igniter, and a flame rod. Complete the Flame Rod Pilot System Test Checklist and make sure to check each step off in the appropriate box as you finish it. This will help you to keep track of your progress.

Flame Rod Pilot System Test Checklist

Step	Procedure	Check
1	Locate the pilot adjustment screw. This is usually a needle valve screw located under a screw cap labeled *pilot adj*.	
2	Turn off main power.	
3	Install the micro-ammeter in series with the flame rod sensor wire to the sensor wire terminal of the control box.	
4	Turn on power, and obtain a pilot only flame by turning off the manual main burner valve or pull the wire labeled *main* at the redundant gas valve. Consult the wiring diagram as required to locate the main gas wire.	
5	Temporarily wire a small toggle switch in series with the main burner wire and the main gas valve. Turn switch off.	
6	Obtain operation of pilot only flame.	
7	Turn on main burner toggle and observe main burner on.	
8	Read flame signal of pilot only in micro-amps. Current = _____ micro-amps	
9	Turn system off.	
10	Remove and clean flame rod with steel wool.	
11	Read flame signal of pilot only in micro-amps. Current = _____ micro-amps	
12	Has the signal changed from Step 8?	
13	Reposition the flame rod to obtain a better contact with the clean blue flame to gain a stronger signal if possible.	
14	Turn the pilot adjustment screw and observe the pilot flame and corresponding micro-amp signal get smaller.	
15	Obtain the smallest flame possible that is capable of proving the pilot flame. Current = _____ micro-amps	
16	Turn on the main burner toggle and observe the main burner light.	
17	Cycle the burner on and off several times and observe the ignition.	
18	Enlarge the pilot as required to provide for smooth main burner ignition. Current = _____ micro-amps	

PILOT TURNDOWN TEST—FLAME IGNITER TYPE

LABORATORY OBJECTIVE
The purpose of this lab is to test the proper operation of a flame igniter type pilot for a gas furnace. Perform this test on any furnace equipped with a pilot and a combination pilot igniter/pilot proving device.

LABORATORY NOTES
A pilot turndown test is a test to determine the smallest possible pilot capable of proving flame presence to the control circuit and lighting the main burner safely. It is performed by lighting the pilot, measuring the pilot flame signal, and observing a series of main burner ignitions using various pilot sizes. Remember that a cold burner is more difficult to ignite and in the performance of this test the burner becomes warmed up; you must allow sufficient cooldown time to perform the final test. A pilot turndown test can also point to other pilot problems.

FUNDAMENTALS OF HVAC/R TEXT REFERENCE
Unit 39

Required Tools and Equipment

Operating gas furnace	Unit 38
Tool kit	Unit 10

SAFETY REQUIREMENTS

A. Never allow gas flow without a flame. If gas is allowed to build up and then suddenly ignite, this can create a serious hazard.

B. If the furnace contains a pilot, always ensure it is properly lit prior to starting the unit.

C. Check all circuits for voltage before doing any service work.

PROCEDURE

STEP 1. Perform this test on any furnace equipped with a pilot and a combination pilot igniter/pilot proving device. Complete the Flame Igniter Pilot System Test Checklist and make sure to check each step off in the appropriate box as you finish it. This will help you to keep track of your progress.

Flame Igniter Pilot System Test Checklist

Step	Procedure	Check
1	Locate the pilot adjustment screw. This is usually a needle valve screw located under a screw cap labeled *pilot adj*.	
2	Turn off main power.	
3	Install an appropriate toggle switch in series with the main gas terminal of the redundant gas valve.	
4	Turn on power and obtain a pilot only flame.	
5	Turn pilot adjustment valve in (clockwise, cw) and observe the pilot flame get smaller.	
6	Continue turning in until the pilot goes out and pilot igniter goes out.	
7	Turn pilot adjustment out (counterclockwise, ccw) and observe the pilot relight.	
8	With the pilot on, turn on the main gas toggle and observe the main gas burner turn on.	
9	With the main burner on, adjust the pilot flame smaller.	
10	Observe the main burner go off when the pilot flame goes out.	
11	Adjust the pilot flame to a size that will light easily and ignite the main burner assembly.	

SEPARATE PILOT GAS PRESSURE REGULATOR

LABORATORY OBJECTIVE

The purpose of this lab is to test the proper operation of a separate pilot gas pressure regulator for a gas furnace. This laboratory worksheet deals with any commercial burner controller utilizing a separate pilot gas pressure regulator, such as a Honeywell commercial RA89F series controller. This procedure will use a plug in flame monitor jack or wire meter in series with the flame rod.

LABORATORY NOTES

A pilot turndown test is a test to determine the smallest possible pilot capable of proving flame presence to the control circuit and lighting the main burner safely. It is performed by lighting the pilot, measuring the pilot flame signal, and observing a series of main burner ignitions using various pilot sizes.

FUNDAMENTALS OF HVAC/R TEXT REFERENCE
Unit 39

Required Tools and Equipment

Operating gas furnace	Unit 38
Tool kit	Unit 10
Micro-ammeter	Unit 11
Plug-in flame monitor	

SAFETY REQUIREMENTS

A. Never allow gas flow without a flame. If gas is allowed to build up and then suddenly ignite, this can create a serious hazard.

B. If the furnace contains a pilot, always ensure it is properly lit prior to starting the unit.

C. Check all circuits for voltage before doing any service work.

PROCEDURE

STEP 1. Perform this test on commercial gas furnace power burners. Complete the Separate Pilot Gas Pressure Regulator Test Checklist and make sure to check each step off in the appropriate box as you finish it. This will help you to keep track of your progress.

Separate Pilot Gas Pressure Regulator Test Checklist

Step	Procedure	Check
1	Turn off power, main gas valve, and pilot gas valve.	
2	Obtain a micro-ammeter and a plug-in flame monitor jack or wire the micro-ammeter in series with the flame rod.	
3	Remove the primary control cover.	
4	Open the pilot gas valve and turn on the power.	
5	Observe the pilot flame only.	
6	Record the pilot micro-amp reading. Current = _____ micro-amps	
7	Decrease the gas pressure by turning the gas pressure regulator out (counterclockwise, ccw). Observe the pilot flame get smaller and the flame signal reduce in current (micro-amps).	
8	Turn the pilot down until the controller relay opens. Current = _____ micro-amps	
9	Increase the pilot flame size until the controller relay opens. Current = _____ micro-amps	
10	Observe the main burner go off when the pilot flame goes out.	
11	Turn on the main burner gas and power.	
12	Observe the main burner ignition.	
13	Adjust the pilot flame as required to obtain smooth burner ignition during cold start. Remember that a cold burner is more difficult to ignite and in the performance of this test the burner becomes warmed up; you must allow sufficient cooldown time to perform the final test.	

CHECK/TEST/REPLACE HOT SURFACE IGNITER

LABORATORY OBJECTIVE

The purpose of this lab is to check, test, and replace a hot surface igniter for a gas furnace.

LABORATORY NOTES

Hot surface ignition and flame proving is the most common method of burner control in modern gas furnaces. The major replacement item in this type of system is the hot surface igniter.

Hot surface igniters are sometimes called glow coils because they glow when energized. An igniter that does not glow is the first indication that it has failed. A typical startup sequence will occur as follows: the thermostat calls for heat, the draft fan comes on, the hot surface igniter glows, the gas valve opens (you can hear the click), the gas ignites, the gas flame reaches the flame rod and proves, and the normal heat mode is in progress. When the surface igniter fails to glow, you hear the click of the gas valve opening but no flame will appear and since the flame does not prove, the gas valve will close.

FUNDAMENTALS OF HVAC/R TEXT REFERENCE
Unit 39

Required Tools and Equipment

Operating gas furnace	Unit 38
Tool kit	Unit 10
Multimeter	Unit 11

SAFETY REQUIREMENTS

A. Check all circuits for voltage before doing any service work.

B. Never allow gas flow without a flame.

PROCEDURE

STEP 1. Familiarize yourself with the gas furnace components. Complete the Gas Furnace Component Identification Checklist and make sure to check each step off in the appropriate box as you finish it. This will help you to keep track of your progress.

Gas Furnace Component Identification Checklist

Step	Procedure	Check
1	Locate the inducer fan and induced draft vent system.	
2	Locate the hot surface igniter.	
3	Locate the flame rod for the flame proving system.	
4	Locate the electronic module or the flame system control.	
5	Locate and write down the terminal on the module that the flame rod wire connects to.	
6	Locate the plug or wire connection from the control system to the hot surface igniter.	

STEP 2. Observe a normal trial for an ignition sequence.

    A. Complete the Normal Ignition Sequence Trial Checklist and make sure to check each step off in the appropriate box as you finish it.

Normal Ignition Sequence Trial Checklist

Step	Procedure	Check
1	Adjust the thermostat to call for heat.	
2	Observe the draft inducer fan come on.	
3	Does the hot surface ignite? (This takes about a minute after the fan comes on.) (circle one) YES NO	
4	If the surface igniter glows, the gas valve should open and the burner flame should light and the furnace should operate normally and the trial is complete. If the surface igniter does not glow and the burner fails to light, proceed to the next step.	
5	If the surface igniter does not glow, did you hear the gas valve click as it opened? (circle one) YES NO	
6	If the surface igniter did not glow, the gas valve clicked open, and the burner did not light, then the surface igniter must be checked.	

7	Make sure all power to the furnace is off. Lock and tag the power panel before removing any parts.	
8	Disconnect the wire nuts from the surface igniter leads or unplug it and test the surface igniter for continuity with a multimeter (refer to Laboratory Worksheet E-1).	
9	If there is continuity, the surface igniter may not be the problem and the voltage to the igniter will need to be verified. This will involve troubleshooting and possibly replacing the module for the flame system control.	
10	If there is no continuity (infinite resistance), then the surface igniter has an open and it will need to be replaced.	
11	Obtain a replacement surface igniter of the same configuration for both the shape of the coil and the connection and replace the defective surface igniter (make sure to supply any shield supplied to protect the coil).	
12	After installing the new surface igniter, turn the power on for the furnace and cycle it through a normal stat mode with the thermostat calling for heat. Observe the main burner ignition and proper operation.	

OIL FIRED FURNACE STARTUP

LABORATORY OBJECTIVE

The purpose of this lab is to demonstrate your ability to go through the necessary sequence for a typical oil furnace startup.

LABORATORY NOTES

You will need to familiarize yourself with oil furnace arrangements and types. Always remember to make a visual check of all system components prior to starting an oil furnace.

FUNDAMENTALS OF HVAC/R **TEXT REFERENCE**
Unit 43

Required Tools and Equipment

Operating forced hot air oil furnace	Unit 42
Tool kit	Unit 10
Multimeters	Unit 11
Clamp-on ammeter	Unit 11

SAFETY REQUIREMENTS

A. Never allow fuel oil to flow into the combustion chamber without a flame. If fuel oil is allowed to build up and then suddenly ignite, this can create a serious hazard.

B. Check all circuits for voltage before doing any service work.

C. Stand on dry nonconductive surfaces when working on live circuits.

D. Never bypass any electrical protective devices.

PROCEDURE

STEP 1. Collect the oil furnace data and fill in the chart.

Oil Furnace Unit Data

Furnace make:		Model number:	Blower motor amperage rating:	
Furnace type (circle one)	Upflow	Counterflow	Basement	Horizontal
Blower type (circle one)	Direct drive	Belt drive		
Blower speed (circle one)	Single speed	Two speed	Three speed	Four speed

STEP 2. Complete the Prestart Checklist and make sure to check each step off in the appropriate box as you finish it. This will help you to keep track of your progress.

Prestart Checklist

Step	Procedure	Check
1	Turn the thermostat down below the room temperature.	
2	Make sure all power to the furnace is off. Lock and tag the power panel before removing any parts.	
3	Check that the vent connector is connected with three screws per joint.	
4	Check that the fuel line is installed properly, with no apparent leaks.	
5	Spin all fans to be sure they are loose and turn freely.	
6	Inspect the combustion area to make sure it is free from debris.	
7	Electrical connections are complete—main power.	
8	Electrical connections are complete—thermostat.	
9	All doors and panels are available and in place.	
10	Thermostat is installed and operating correctly.	
11	Fuel oil is present in tank.	
12	Barometric damper is installed and swinging freely.	

STEP 3. After finishing the Prestart Checklist you are now ready to start the furnace and check its operation. You will follow the same procedure of working through a check list. This will help you to keep track of your progress. Complete the Startup Checklist and make sure to check each step off in the appropriate box as you finish it.

Startup Checklist

Step	Procedure	Check
1	Check the main fuse or breaker and the amperage rating and make sure it is the correct rating for the furnace.	
2	Measure the incoming supply voltage. Refer to Laboratory Worksheet E-1 for the proper procedure for testing the circuit.	
3	If Steps 1 and 2 have been completed and are satisfactory, then turn on the supply power for the furnace.	
4	Set the thermostat to FAN ON to obtain the fan only operation and observe that the blower is operating. If the thermostat is not so equipped, then consult with your instructor.	
5	Turn the fan off and the thermostat to heat.	
6	Adjust the thermostat setting to 10°F above the room temperature.	
7	Observe the burner come on.	
8	Observe the fan come on after a reasonable warm-up time.	
9	Use a clamp-on ammeter to measure the blower motor amperage and compare this value to its rating. Refer to Laboratory Worksheet E-1 for the proper procedure for using electrical meters.	
10	Turn the down the thermostat and observe the burner shut off, and then the fan shut off, in that order.	
11	Turn on and off for several ignitions.	
12	The flame should be orange/white in color, uniform in shape, and within the combustion chamber, quiet and quick in ignition and extinction, and with no drip from the burner tip.	

OIL BURNER TUNE-UP

LABORATORY OBJECTIVE

The purpose of this lab is to demonstrate your ability to conduct the proper burner tune-up for an oil fired furnace.

LABORATORY NOTES

You will need to familiarize yourself with oil furnace arrangements and types. Always remember to make a visual check of all system components prior to starting an oil furnace.

FUNDAMENTALS OF HVAC/R TEXT REFERENCE
Unit 43

Required Tools and Equipment

Operating oil furnace	Unit 42
Tool kit	Unit 10
Multimeters	Unit 11
Clamp-on ammeter	Unit 11

SAFETY REQUIREMENTS

A. Never allow fuel oil to flow into the combustion chamber without a flame. If fuel oil is allowed to build up and then suddenly ignite, this can create a serious hazard.

B. Check all circuits for voltage before doing any service work.

C. Stand on dry nonconductive surfaces when working on live circuits.

D. Never bypass any electrical protective devices.

PROCEDURE

STEP 1. Collect the oil fired furnace data and fill in the chart.

Oill Fired Furnace Unit Data

Furnace make:		Model number:	Blower motor amperage rating:	
Furnace type (circle one)	Upflow	Counterflow	Basement	Horizontal
Burner make:			Burner model number:	
Recommended nozzle data	Gallons per hour (gph):		Spray pattern:	Spray angle:

STEP 2. Complete the Prestart Checklist and the Startup Checklist as outlined in the procedure provided in Lab OH-1. After finishing the Startup Checklist, shut down the furnace. Make sure all power to the furnace is off. Lock and tag the power panel before removing any parts. Complete the Oil Burner Inspection and Tune-Up Checklist.

Oil Burner Inspection and Tune-Up Checklist

Step	Procedure	Check
1	Make sure all power to the furnace is off. Lock and tag the power panel before removing any parts.	
2	Open or remove transformer.	
3	Remove nozzle assembly.	
4	Remove and clean electrodes.	
5	Remove nozzle and record nozzle data. Make _____ gph _____ Angle _____ Pattern _____	
6	Replace the nozzle with a new nozzle that matches the data specifications.	
7	Install and adjust electrodes.	
8	Position electrodes to manufacturer recommended setting. (Typically these dimensions are approximately 1/2 in above, 1/16 in forward, and 1/8 in apart.)	
9	Remove and clean cad cell.	

Step	Procedure	Check
10	Remove and record resistance of cad cell while the face of the cad cell is covered (refer to Laboratory Worksheet E-1). Resistance = _____ Ohms	
11	Remove and record resistance of cad cell while the face of the cad cell is exposed to room light. Resistance = _____ Ohms	
12	Install nozzle assembly.	
13	Slide nozzle assembly to midpoint of forward/backward adjustment.	
14	Swing up or reinstall ignition transformer.	
15	Make sure that the transformer contact springs touch the electrodes as the transformer is positioned.	
16	Secure the transformer in place with at least one screw for testing.	
17	Turn on power to the furnace and burner and observe ignition. Do not get too close as oil burners can puff back.	
18	Observe a normal continuous burn for 1 min.	
19	Adjust the air shutter back and forth slowly. Close to see smoke and then open until smoky flame disappears.	
20	Slide nozzle assembly forward and back until the flame is quiet and no longer jagged and smoky.	
21	Refer to Laboratory Worksheet OH-3 for final burner adjustments.	

FINAL OIL BURNER ADJUSTMENT

LABORATORY OBJECTIVE

The purpose of this lab is to demonstrate your ability to conduct the proper final burner adjustments after a tune-up for an oil fired furnace.

LABORATORY NOTES

You will need to familiarize yourself with oil furnace arrangements and types. Always remember to make a visual check of all system components prior to starting an oil furnace.

FUNDAMENTALS OF HVAC/R TEXT REFERENCE
Unit 43

Required Tools and Equipment

Operating oil furnace	Unit 42
Tool kit	Unit 10
Draft gauge	Unit 13
Smoke gun	. Unit 13
CO analyzer	Unit 13

SAFETY REQUIREMENTS

A. Never allow fuel oil to flow into the combustion chamber without a flame. If fuel oil is allowed to build up and then suddenly ignite, this can create a serious hazard.

B. Check all circuits for voltage before doing any service work.

C. Stand on dry nonconductive surfaces when working on live circuits.

D. Never bypass any electrical protective devices.

PROCEDURE

STEP 1. Complete the Prestart Checklist and the Startup Checklist as outlined in the procedure provided in Laboratory Worksheet OH-1. Complete the Oil Burner Inspection and Startup Checklist as outlined in Laboratory Worksheet OH-2. Obtain a draft gauge, smoke gun, and CO analyzer to perform final settings on the oil burner.

Final Burner Adjustment Checklist

Step	Procedure	Check
1	With the burner operating, use the draft gauge and adjust the barometric damper to obtain a –0.1 over fire draft.	
2	Measure and record the draft at the breach of the burner. A restriction of greater than –0.4 through the heat exchanger indicates the heat exchanger may be plugged with soot and needs to be cleaned. Final measured draft at breach _____	
3	If necessary, inspect and clean the fire side of the heat exchanger. Make sure all power to the furnace is off. Lock and tag the power panel before removing any parts.	
4	Once cleaning is complete, restart the burner and perform a smoke spot test with the smoke gun. The reading should be a #0.	
5	Install the CO meter into the breach of the furnace.	
6	Readjust the air shutter to obtain 10% CO_2.	
7	Turn the burner on and off, observing several ignitions.	
8	Verify a quick, clean, quiet ignition.	
9	Remove and put away all test equipment and tools. Install all panels and doors.	

Name _____

Date _____

Instructor's OK ☐

OIL FURNACE PREVENTIVE MAINTENANCE

LABORATORY OBJECTIVE

The purpose of this lab is to demonstrate your ability to conduct the proper preventive maintenance for an oil fired furnace.

LABORATORY NOTES

You will need to familiarize yourself with oil furnace arrangements and types. Always remember to make the visual check of all system components prior to starting an oil furnace.

***FUNDAMENTALS OF HVAC/R* TEXT REFERENCE**
Unit 43

Required Tools and Equipment

Operating oil furnace	Unit 42
Tool kit	Unit 10
Multimeters	Unit 11
Clamp-on ammeter	Unit 11
Oil pressure gauge	Unit 13
Vacuum	

SAFETY REQUIREMENTS

    A. Never allow fuel oil to flow into the combustion chamber without a flame. If fuel oil is allowed to build up and then suddenly ignite, this can create a serious hazard.

    B. Check all circuits for voltage before doing any service work.

    C. Stand on dry nonconductive surfaces when working on live circuits.

    D. Never bypass any electrical protective devices.

PROCEDURE

STEP 1. Collect the oil fired furnace data and fill in the chart.

Oil Fired Furnace Unit Data

Furnace make:		Model number:	Blower motor amperage rating:	
Furnace type (circle one)	Upflow	Counterflow	Basement	Horizontal
Burner make:		Oil pressure:	Burner model number:	
Recommended nozzle data	Gallons per hour (gph):		Spray pattern:	Spray angle:

STEP 2. Complete the Prestart Checklist and the Startup Checklist as outlined in the procedure provided in Laboratory Worksheet OH-1. After finishing the Startup Checklist, shut down the furnace. Make sure all power to the furnace is off. Lock and tag the power panel before removing any parts. Complete the Blower Maintenance Checklist.

Blower Maintenance Checklist

Step	Procedure	Check
1	Make sure that the power to the blower motor is secured. Verify this with a multimeter voltage test. Refer to Laboratory Worksheet E-1.	
2	Remove the wires from the blower motor at an accessible location.	
3	Remove screws or bolts securing the blower assembly.	
4	Inspect belt for cracks and signs of wear.	
5	Inspect pulleys for wear, grooving, and alignment.	
6	Spin blower by hand and observe pulleys turn.	
7	Inspect for pulley wobble and alignment.	
8	Listen for bearing noise, drag, or movement.	
9	Inspect blower shaft for signs of wear.	
10	Clean blower and motor with air pressure, brushes, scrapers, and cleaning solution as required.	
11	Oil motor and blower bearings as required.	
12	Reassemble blower, taking care to check pulley alignment and correct belt tension.	
13	Check the condition of the heat exchanger before reinstalling the blower.	

STEP 3. After finishing the Blower Maintenance Checklist, you can perform the necessary oil burner maintenance. Again, make sure all power to the furnace is off. Lock and tag the power panel before removing any parts. Complete the Oil Burner Maintenance Checklist.

Oil Burner Maintenance Checklist

Step	Procedure	Check
1	Make sure all power to the furnace is off. Lock and tag the power panel before removing any parts.	
2	Remove and clean nozzle assembly.	
3	Replace nozzle with a manufacturer recommended nozzle.	
4	Place the old nozzle into a small zipper bag with the new nozzle box. Write your name on the bag, date it, and leave it with the furnace. The bag will keep it from smelling and then on the next service visit, it will be evident what was put in and what was taken out.	
5	Position electrodes to manufacturer's recommended setting. (Typically these dimensions are approximately 1/2 in above, 1/16 in forward, and 1/8 in apart.)	
6	Remove the vent connector, barometric damper, and flue baffles or cleanout plugs.	
7	Remove or swing out burner assembly.	
8	Vacuum and brush all soot, rust, and solid particles from the fire side of the heat exchanger. Be careful not to damage the combustion chamber refractory.	
9	Insert a light into the combustion chamber and if possible turn off the lights in the furnace room.	
10	With the light in as far back as possible, inspect the heat exchanger from the fan side and the plenum side for holes, light, or leaky gaskets.	
11	After inspecting the heat exchanger reinstall the blower.	
12	Reinstall the nozzle assembly while the burner is still swung away from the furnace.	
13	After the burner assembly is back in place, the fuel oil system can be checked. Disconnect the supply line from the pump to the burner assembly and install an oil pressure gauge on the pump supply line.	
14	Remove the lockout tag and turn on the burner and check the initial fuel oil pressure. Fuel oil pressure = _____	
15	Adjust to the manufacturer recommended fuel oil pressure (typically 100 psig). Fuel oil pressure = _____	
16	Observe the burner turn off due to flame failure.	
17	Observe oil pressure gauge holding 85 psig quick cutoff pressure.	

(Continued)

Oil Burner Maintenance Checklist (*Continued*)

Step	Procedure	Check
18	Reinstall the original oil supply line to the nozzle assembly.	
19	Check the fan control settings. The fan should turn on at a supply air temperature of approximately 90°F and off at 135°F.	
20	The maximum temperature limit setting can be checked by running the burner with the fan off and at approximately 200°F, the burner should cycle off.	
21	After all maintenance is complete, turn the burner on and off, observing several ignitions. Verify a quick, clean, quiet ignition. Remove and put away all test equipment and tools. Install all panels and doors.	

OIL FURNACE STORAGE TANK MAINTENANCE

LABORATORY OBJECTIVE

The purpose of this lab is to demonstrate your ability to conduct the proper preventive maintenance for an oil fired furnace oil storage tank.

LABORATORY NOTES

You will need to familiarize yourself with oil furnace arrangements and types. Always remember to make a visual check of all system components prior to starting an oil furnace.

***FUNDAMENTALS OF HVAC/R* TEXT REFERENCE**

Unit 43

Required Tools and Equipment

Oil furnace tank and operating oil furnace	Unit 42
Tool kit	Unit 10
Oil adsorbent pads	
Large empty drip pan	

SAFETY REQUIREMENTS

A. A full oil storage tank will normally hold 250 gal of fuel oil or more. Always make sure the shutoff valve is closed before changing the tank filter to prevent the possibility of a large oil spill.

B. It is good practice to have oil adsorbent pads and a large drip pan available to collect any minor oil spills that may occur when changing oil tank filters.

PROCEDURE

STEP 1. Complete the Oil Furnace Storage Tank Maintenance Checklist and make sure to check each step off in the appropriate box as you finish it. This will help you to keep track of your progress.

Oil Furnace Storage Tank Maintenance Checklist

Step	Procedure	Check
1	Make sure all power to the furnace is off. Lock and tag the power panel before removing any parts.	
2	Visually check the oil storage tank for leaks. Older tanks will rust from the inside out and leaks may develop.	
3	Close the tank shut-off valve located between the tank and the filter assembly.	
4	Place an adsorbent pad beneath the filter assembly and place the large empty drip pan beneath the filter assembly to catch any dripping fuel oil.	
5	Carefully remove the filter assembly and check for water. Water is heavier than the fuel oil and will settle to the bottom of the fuel tank.	
6	If water is present, additional fuel may need to be drained from the tank.	
7	All fuel removed along with the old filter element, adsorbent pads, and any rags, must be removed and disposed of properly. Never leave any of these at the job site.	
8	Always replace the filter assembly gaskets with new ones.	
9	After the new filter assembly is in place and tight, the fuel tank shut-off valve can be opened and the air bled through the vent screw located on the top of the filter assembly. Vent air from the assembly until fuel comes from the vent screw. Catch any dripping fuel in the drip pan.	
10	Wipe the filter assembly dry and check for any fuel leaks.	
11	After the new filter as been installed, the fuel line from the tank to the oil pump must be purged of air.	
12	Locate the purge valve on the fuel oil pump located on the furnace burner.	
13	Place the drip pan beneath the oil pump purge valve.	
14	Remove the lock on the furnace power and start the burner. Slowly open the oil pump purge valve and fuel oil will squirt out to collect into the drip pan. Allow the oil to flow from the purge valve until all of the air has been purged from the line, then close the purge valve.	
15	If air remains in the fuel line, the burner will not light normally and go out.	
16	After purging the air from the oil line, turn the burner on and off, observing several ignitions. Verify a quick clean quiet ignition.	
17	All fuel removed along with the old filter element, adsorbent pads, and any rags, must be removed and disposed of properly. Never leave any of these at the job site.	

OIL FURNACE TWO PIPE CONVERSION

LABORATORY OBJECTIVE

The purpose of this lab is to demonstrate your ability to convert a one pipe oil system to a two pipe oil system for an oil fired furnace.

LABORATORY NOTES

A one pipe oil system refers to the oil line from the tank to the burner being a single pipe, usually 3/8 in OD copper. This system is recommended for no more than 2 ft of vertical lift from the oil level in the tank to the burner pump. Any time the tank runs out of oil, the air must be manually bled from the line at the pump.

A two pipe conversion involves running a second line back to the tank and installing a bypass plug within the fuel oil pump. The second line at the tank needs to go all the way to the bottom of the tank. If oil entered the top of the tank and fell to the bottom, you would hear the oil fall and splash. Frequently a tank duplex fitting is used on two pipe systems. The fitting is installed in the top of the tank and has two fittings that will allow a 3/8 in line to be pushed through to the bottom of the tank and then pulled up 3 to 4 in. This is the correct position to install both the supply line and the return line. This reduces the chance of pulling sludge or water off the bottom of the tank.

FUNDAMENTALS OF HVAC/R TEXT REFERENCE
Unit 44

Required Tools and Equipment

Oil furnace tank and operating oil furnace	Unit 42
Tool kit	Unit 10
Oil adsorbent pads	
Large empty drip pan	
Tank duplex fitting	

SAFETY REQUIREMENTS

A. A full oil storage tank will normally hold 250 gal of fuel oil or more. Always make sure the shut-off valve is closed to prevent the possibility of a large oil spill.

B. It is good practice to have oil adsorbent pads and a large drip pan available to collect any minor oil spills that may occur.

PROCEDURE

STEP 1. Complete the System Inspection Checklist and make sure to check each step off in the appropriate box as you finish it. This will help you to keep track of your progress.

System Inspection Checklist

Step	Procedure	Check
1	Identify the current fuel oil piping system. (circle one) One Pipe Two Pipe	
2	Measure the lift from the lowest possible operating oil level to the burner pump fitting. Lift = _____ ft	
3	Is there a bypass plug at the pump? Every new oil burner comes with a bypass plug in a cloth bag generally attached to the pump with string. Is it still there? (circle one) Yes No	
4	If yes, you have the plug you need. If no, you will need to get one.	
5	Obtain a bypass plug as required.	
6	Obtain a sufficient length of 3/8 in OD copper tubing and the required brass fittings to make connection at the pump.	

STEP 2. After completing the System Inspection Checklist you can prepare to convert to a two pipe system. Complete the Two Pipe System Conversion Checklist and make sure to check each step off in the appropriate box as you finish it.

Two Pipe System Conversion Checklist

Step	Procedure	Check
1	Make sure all power to the furnace is off. Lock and tag the power panel before removing any parts.	
2	Close the tank shut-off valve.	
3	Disconnect the existing one line pipe. Place an adsorbent pad large empty drip pan beneath the connection to catch any dripping fuel oil.	
4	Loosen and remove the bolts holding the fuel oil pump in position.	
5	Inspect the pump fitting opening for the return line to the tank (pumps are generally labeled for openings and plug location).	
6	Inspect the pump for the location of the bypass plug. This is generally a 1/8 in or 1/16 in female pipe thread inside the return line opening. Refer to the manufacturer's data for the pump as required.	
7	Hold the pump at an upward angle. Use an Allen wrench of sufficient length to install and tighten the bypass plug.	
8	Install the copper tubing to the 3/8 in flare fitting at the pump.	
9	Mount the pump back onto the burner housing.	
10	Run the 3/8 in line from the pump outlet into the top of the tank and down into the tank at 3 to 4 in from the bottom.	
11	Snug the fitting to hold the line in place.	
12	Place supports for the line at appropriate locations from the pump to the tank.	
13	After the line has been installed, the tank shut-off valve may be opened, and the power restored to the furnace. The burner should start and the fuel line should self-bleed itself of air with the bypass oil returning to the fuel tank.	
14	All fuel removed along with the adsorbent pads, and any rags, must be removed and disposed of properly. Never leave any of these at the job site.	

INSTALL A REPLACEMENT FUEL OIL PUMP

LABORATORY OBJECTIVE

The purpose of this lab is to demonstrate your ability to install a replacement fuel oil pump for an oil fired furnace.

LABORATORY NOTES

The pump on an oil burner is one of the most important parts of the burner assembly. Its job is to pull the fuel oil from the tank and deliver it to the burner nozzle at the recommended supply pressure (typically 100 or 140 psig). Always check the burner name plate to verify the manufacturer's recommended fuel oil supply pressure. Minor adjustments can be made; however, if the pump is worn, it must be replaced.

There are dozens of different pumps available. This is because there is more than one pump manufacturer, several major burner manufacturers, and different types of burner configurations. Not every wholesale house has every pump. To find a replacement pump, begin by contacting the dealer of the burner that you are working on. You will need the furnace and burner model numbers and serial numbers.

FUNDAMENTALS OF HVAC/R TEXT REFERENCE

Unit 44

Required Tools and Equipment

Oil furnace tank and operating oil furnace	Unit 42
Tool kit	Unit 10
Oil adsorbent pads	
Large empty drip pan	

SAFETY REQUIREMENTS

It is good practice to have oil adsorbent pads and a large drip pan available to collect any minor oil spills that may occur.

PROCEDURE

STEP 1. Collect the oil fired furnace data and fill in the chart.

Oil Fired Furnace Unit Data

Furnace make:	Model number:	Serial number:
Burner make:	Model number:	Serial number:
Pump make:	Model number:	Serial number:

STEP 2. Complete the System Inspection Checklist and make sure to check each step off in the appropriate box as you finish it. This will help you to keep track of your progress.

System Inspection Checklist

Step	Procedure	Check
1	Identify the current fuel oil piping system. (circle one) One Pipe Two Pipe	
2	Measure the lift from the lowest possible operating oil level to the burner pump fitting. Lift = _____ft	
3	If the lift is greater than 2 ft on a one pipe system, we will need to convert to a two pipe system (see Laboratory Worksheet OH-6).	

STEP 3. After completing the System Inspection Checklist you can prepare to replace the fuel oil pump. Complete the Fuel Pump Replacement Checklist and make sure to check each step off in the appropriate box as you finish it.

Fuel Pump Replacement Checklist

Step	Procedure	Check
1	Make sure all power to the furnace is off. Lock and tag the power panel before removing any parts.	
2	Close the tank shut-off valve.	
3	Disconnect the existing oil line or lines. Place an adsorbent pad and a large empty drip pan beneath the connection to catch any dripping fuel oil.	
4	Loosen and remove the bolts holding the fuel oil pump in position.	
5	Change fittings to the new pump (change to a two pipe system if required; see Laboratory Worksheet OH-6).	
6	Bolt new pump into position.	
7	After the replacement fuel oil pump has been installed, the tank shut-off valve may be opened, and the power may be restored to the furnace.	
8	On a two pipe system, the burner should start and the fuel line should self-bleed itself of air with the bypass oil returning to the fuel tank.	
9	On a one pipe system, after the replacement pump has been installed, the fuel line from the tank to the oil pump must be purged of air.	
10	Locate the purge valve on the fuel oil pump located on the furnace burner.	
11	Place the drip pan beneath the oil pump purge valve.	
12	Start the burner and slowly open the oil pump purge valve and fuel oil will squirt out to collect into the drip pan. Allow the oil to flow from the purge valve until all of the air has been purged from the line, then close the purge valve.	
13	If air remains in the fuel line, the burner will not light normally and it will go out.	
14	After purging the air from the oil line, turn the burner on and off, observing several ignitions. Verify a quick, clean, quiet ignition.	
15	All fuel removed along with the old filter element, adsorbent pads, and any rags, must be removed and disposed of properly. Never leave any of these at the job site.	

CHECK/TEST A CAD CELL OIL BURNER PRIMARY CONTROL

LABORATORY OBJECTIVE

The purpose of this lab is to demonstrate your ability to test for the proper operation of a cadmium sulfide cell (cad cell) for an oil fired furnace.

LABORATORY NOTES

The cadmium sulfide cell (cad cell) of an oil burner primary control system proves the presence of an oil flame by observing the visible light from the flame. The cad cell's electrical resistance is greatly reduced in the presence of light. The resistance must be high to enable the primary to initiate a trial for ignition, also called a dark start function. Once a flame is established, the light from the flame causes the cad cell's resistance to drop and the flame will continue. During the trial ignition, a safety switch heater is energized that will open the safety switch contacts and lock out the burner if the flame is not proved within the trial for ignition time, usually 30, 45, or 60 s. This heater must cool off before the burner can be reset manually and started again. When a flame is established and proved by the cad cell, the safety switch heater is deenergized and the contacts remain closed.

FUNDAMENTALS OF HVAC/R TEXT REFERENCE

Unit 43

Required Tools and Equipment

Operating oil furnace	Unit 42
Tool kit	Unit 10
Ohmmeter	Unit 11
Clamp-on ammeter	Unit 11
1,200 ohm resistor	
Timer for timing seconds	

SAFETY REQUIREMENTS

A. Never allow fuel oil to flow into the combustion chamber without a flame. If fuel oil is allowed to build up and then suddenly ignite, this can create a serious hazard.

B. Check all circuits for voltage before doing any service work.

C. Stand on dry nonconductive surfaces when working on live circuits.

D. Never bypass any electrical protective devices.

PROCEDURE

STEP 1. Collect the oil fired furnace data and fill in the chart.

Oil Fired Furnace Unit Data

Furnace make:		Model number:	Blower motor amperage rating:	
Furnace type (circle one)	Upflow	Counterflow	Basement	Horizontal
Burner make:			Burner model number:	
Recommended nozzle data	Gallons per hour (gph):		Spray pattern:	Spray angle:

STEP 2. Complete the Prestart Checklist and the Startup Checklist as outlined in the procedure provided in Laboratory Worksheet OH-1. After finishing the Startup Checklist, shut down the furnace. Make sure all power to the furnace is off. Lock and tag the power panel before removing any parts. Complete the Cad Cell Test Checklist.

Cad Cell Test Checklist

Step	Procedure	Check
1	Make sure all power to the furnace is off. Lock and tag the power panel before removing any parts.	
2	Open or remove transformer.	
3	Locate and unplug cad cell from plug mount.	
4	Inspect and wipe clean the lens cover of the cad cell.	
5	Remove and record resistance of cad cell while the face of the cad cell is covered (refer to Laboratory Worksheet E-1). Resistance = _____ ohms	
6	Remove and record resistance of cad cell while the face of the cad cell is exposed to room light. Resistance = _____ ohms	
7	Insert the cad cell into the plug assembly.	
8	Locate yellow wires from the cad cell mount at the primary control terminals F and F.	

Step	Procedure	Check
9	Remove cad cell wires from F and F and connect to the ohmmeter.	
10	Swing transformer slowly closed with the ohmmeter still connected.	
11	Turn on power to the furnace and burner and observe ignition. Do not get too close as oil burners can puff back.	
12	Read and record the resistence of the cad cell exposed to a normal flame. Resistance = _____ ohms	
13	The flame will shut down and the burner will lock out within 30 s because the cad cell is not connected to the primary control.	
14	Obtain a 1,200 Ω resistor and connect one end of the resistor to one F terminal of the primary control.	
15	Push the reset button. Wait a minimum of 2 min cooldown time.	
16	Observe the burner start and the flame ignite.	
17	Carefully connect the second wire of the 1,200 ohm resistor to the other F terminal of the primary control within the 30 s trial for ignition time.	
18	Observe the flame continue for another 5 min.	
19	You will time how long it takes in seconds for the flame to shut down and the burner to lock out when you remove one lead from the resistor, which will simulate a flame failure. Running time after removing resistor = _____ s	
20	After completing this test, make sure all power to the furnace is off. Lock and tag the power panel before removing any parts.	
21	Remove the resistor and ohmmeter and reconnect the cad cell.	
22	After the cad cell has been reconnected, turn the burner on and off, observing several ignitions. Verify a quick, clean, quiet ignition.	

INSTALL A REPLACEMENT FUEL OIL BURNER

LABORATORY OBJECTIVE

The purpose of this lab is to demonstrate your ability to install a replacement fuel oil burner for an oil fired furnace.

LABORATORY NOTES

Many oil furnaces are constructed of heavy metal and are quite durable. It is not uncommon for the burner to become worn out while the basic furnace is still in good condition. In such cases the entire burner can be replaced. This has the advantage of a matched nozzle assembly and flame cone along with a new pump, motor, transformer, and primary control. This type of replacement is less expensive, faster, and easier than installing an entirely new furnace.

FUNDAMENTALS OF HVAC/R TEXT REFERENCE

Unit 44

Required Tools and Equipment

Oil furnace tank and operating oil furnace	Unit 42
Tool kit	Unit 10
Oil adsorbent pads	
Large empty drip pan	
Vacuum	

SAFETY REQUIREMENTS

It is good practice to have oil adsorbent pads and a large drip pan available to collect any minor oil spills that may occur.

PROCEDURE

STEP 1. Collect the oil fired furnace data and fill in the chart.

Oil Fired Furnace Unit Data

Furnace make:	**Model number:**	**Serial number:**
Existing burner make:	**Model number:**	**Serial number:**
New burner make:	**Model number:**	**Serial number:**
Length of blast tube required:		
Identify piping system: (circle one)	One pipe	Two pipe
New combustion chamber? (circle one)	Yes	No
New thermostat? (circle one)	Yes	No

STEP 2. After filling in the Oil Fired Furnace Unit Data chart you can prepare to remove the old burner assembly. Complete the Burner Removal Checklist and make sure to check each step off in the appropriate box as you finish it.

Burner Removal Checklist

Step	Procedure	Check
1	Make sure all power to the furnace is off. Lock and tag the power panel before removing any parts.	
2	Close the fuel tank shut-off valve.	
3	Disconnect the existing oil line or lines. Place an adsorbent pad and a large empty drip pan beneath the connection to catch any dripping fuel oil.	
4	Install 3/8 in flare plugs in both fuel lines to prevent oil leakage.	
5	Carefully bend the fuel lines out of the way. You will reuse the same lines if possible.	
6	Disconnect the main power and thermostat wire from the burner.	
7	Remove the mounting bolts holding the burner assembly in place.	
8	Remove the mounting plate from the front of the furnace.	
9	Use a vacuum to clean any debris from the combustion chamber area. Do not damage the combustion chamber refractory.	
10	Inspect the combustion chamber for any signs of cracks or deterioration. Replace as required.	

STEP 3. After removing the old burner assembly, you may prepare to install the new burner assembly. Complete the Burner Installation Checklist and make sure to check each step off in the appropriate box as you finish it.

Burner Installation Checklist

Step	Procedure	Check
1	Install new combustion chamber and components if required.	
2	Hold the burner mounting plate in position and measure the distance to the combustion chamber. Distance = _____	
3	Measure the length of the blast tube on the new burner. Length = _____	
4	Exchange the blast tube on the burner if the length is not a match.	
5	Check/install the nozzle for correct flow rate (gph), angle, and pattern.	
6	Check and adjust electrode position (refer to Laboratory Worksheet OH-2).	
7	Bolt new mounting plate to furnace.	
8	Bolt new burner with blast tube onto mounting plate.	
9	Connect oil lines to burner. Install bypass plug for two pipe systems.	
10	Replace oil filter in fuel supply line and bleed air from line (refer to Laboratory Worksheet OH-5).	
11	Perform final burner adjustments (refer to Laboratory Worksheet OH-3).	

ELECTRIC FURNACE STARTUP

LABORATORY OBJECTIVE
The purpose of this lab is to demonstrate your ability to go through the necessary sequence for a typical electric furnace startup.

LABORATORY NOTES
A typical electric furnace will have three stages of electric heat, usually 3 or 5 kWh each. One kWh would be 3,400 Btu/hr and pull 41.6 amps at 240 V. A 3 kWh heater would be three times that or 12.5 amps and a 5 kWh heater would be 20.8 amps. Three 5 kW heaters would pull over 60 amps, too much to just turn on. Power companies and some codes require electric furnaces to be equipped with a sequencer that is a time delay device, and multiple contactors. This is installed in the furnace and not part of the thermostat. Even with a single-stage heat-only furnace, the electric heat elements come on one at a time, spaced apart by a few seconds at least. Supply air temperatures of lower than 120°F can feel rather cool and airflow should be reduced to keep the air temperature to a comfortable level.

FUNDAMENTALS OF HVAC/R TEXT REFERENCE
Unit 47

Required Tools and Equipment

Operating electric furnace	Unit 46
Tool kit	Unit 10
Multimeter	Unit 11
Clamp-on ammeter	Unit 11
Temperature sensor	Unit 12

SAFETY REQUIREMENTS
    A. Check all circuits for voltage before doing any service work.

    B. Stand on dry nonconductive surfaces when working on live circuits.

    C. Never bypass any electrical protective devices.

PROCEDURE

STEP 1. Collect the electric furnace data and fill in the chart.

Electric Furnace Unit Data

Furnace make:		Model number:		
Electrical data	Voltage:	Phase:	Amperage:	kW:
Blower type (circle one)	Direct drive	Belt drive		
Blower speed (circle one)	Single speed	Two speed	Three speed	Four speed
System type (circle one)	Heat only	Heat pump and electric backup		
Electric heaters	1st kW:	2nd kW:	3rd kW:	Total kW:

STEP 2. Complete the Prestart Checklist and make sure to check each step off in the appropriate box as you finish it. This will help you to keep track of your progress.

Prestart Checklist

Step	Procedure	Check
1	Make sure all power to the furnace is off. Lock and tag the power panel before removing any parts.	
2	Check all electrical connections for tightness.	
3	Spin all fans to be sure they are loose and turn freely.	
4	Check to make sure all airflow passages are unobstructed.	
5	Make sure all doors and panels are available and are in place.	
6	Check that the thermostat is installed and operating correctly.	

STEP 3. After finishing the Prestart Checklist you are now ready to start the furnace and check its operation. You will follow the same procedure of working through a checklist. This will help you to keep track of your progress. Complete the Startup Checklist and make sure to check each step off in the appropriate box as you finish it.

Startup Checklist

Step	Procedure	Check
1	Check the main fuse or breaker and the amperage rating and make sure it is the correct rating for the furnace.	
2	Measure the incoming supply voltage; refer to Laboratory Worksheet E-1 for the proper procedure for testing the circuit.	
3	If Steps 1 and 2 have been completed and are satisfactory, then turn on the supply power for the furnace.	
4	Set the thermostat to FAN ON to obtain the fan only operation and observe that the blower is operating. If the thermostat is not so equipped then consult with your instructor.	
5	Turn the fan off and turn the thermostat to heat.	
6	Adjust the thermostat setting to 10°F above the room temperature.	
7	Observe the fan start and the heater banks come on.	
8	Use the clamp-on ammeter to measure and record the amperage as the sequence brings on electric banks. 1st = _____ 2nd = _____ 3rd = _____	
9	Obtain normal heating operation.	
10	Measure and record temperatures. Discharge air temperature = _____ Room air temperature = _____	
11	Calculate temperature rise. Discharge air temperature – room air temperature = temperature rise Temperature rise = _____	
12	Turn the thermostat down and observe the heaters come off.	
13	Observe the blower stop 3 min after the heaters come off.	

CALCULATE AIRFLOW BY TEMPERATURE RISE

LABORATORY OBJECTIVE

The purpose of this lab is to demonstrate your ability to calculate airflow by the temperature rise methods through a typical electric furnace.

LABORATORY NOTES

Electric heaters are nearly 100% efficient and allow for a very accurate measurement of the amount of heat produced. The heaters are located within the duct or furnace and there is no heat lost going up the chimney. We can calculate the airflow by the temperature rise method very accurately. The important thing is to measure the voltage, amperage, and temperatures as accurately as possible. The blower amperage must also be kept separate and not added to the heater amperage.

FUNDAMENTALS OF HVAC/R TEXT REFERENCE
Unit 47

Required Tools and Equipment

Operating electric furnace	Unit 46
Tool kit	Unit 10
Multimeter	Unit 11
Clamp-on ammeter	Unit 11
Temperature sensor	Unit 12

SAFETY REQUIREMENTS

A. Check all circuits for voltage before doing any service work.

B. Stand on dry nonconductive surfaces when working on live circuits.

C. Never bypass any electrical protective devices.

PROCEDURE

STEP 1. Collect the electric furnace data and fill in the chart.

Electric Furnace Unit Data

Furnace make:			Model number:		
Electrical data	Voltage:	Phase:		Amperage:	kW:
Blower type (circle one)	Direct drive	Belt drive			
Blower speed (circle one)	Single speed	Two speed		Three speed	Four speed
System type (circle one)	Heat only	Heat pump and electric backup			
Electric heaters	1st kW:	2nd kW:		3rd kW:	Total kW:

STEP 2. Complete the Prestart Checklist and make sure to check each step off in the appropriate box as you finish it. This will help you to keep track of your progress.

Prestart Checklist

Step	Procedure	Check
1	Make sure all power to the furnace is off. Lock and tag the power panel before removing any parts.	
2	Check all electrical connections for tightness.	
3	Spin all fans to be sure they are loose and turn freely.	
4	Check to make sure all airflow passages are unobstructed.	
5	Check that all doors and panels are available and in place.	
6	Check that the thermostat is installed and operating correctly.	
7	Allocate and count each heater element contactor. How many are there and what is the kW rating? _____ of _____ kWh	

Step 3. After finishing the Prestart Checklist you are now ready to start the furnace and check its operation. You will follow the same procedure of working through a checklist. This will help you to keep track of your progress. Complete the Startup Checklist and make sure to check each step off in the appropriate box as you finish it.

Startup Checklist

Step	Procedure	Check
1	Check the main fuse or breaker and the amperage rating and make sure it is the correct rating for the furnace.	
2	Measure the incoming supply voltage; refer to Laboratory Worksheet E-1 for the proper procedure for testing the circuit.	
3	If steps 1 and 2 have been completed and are satisfactory, then turn on the supply power for the furnace.	
4	Set the thermostat to FAN ON to obtain the fan only operation and observe that the blower is operating. If the thermostat is not so equipped then consult with your instructor.	
5	Turn the fan off and the thermostat to HEAT.	
6	Adjust the thermostat setting to 10°F above the room temperature.	
7	Observe the fan start and the heater banks come on.	
8	Use the clamp-on ammeter to measure and record the amperage as the sequence brings on electric banks. 1st = _____ 2nd = _____ 3rd = _____	
9	Obtain normal heating operation.	
10	Measure and record temperatures. Discharge air temperature = _____ Room air temperature = _____	
11	Calculate temperature rise. Discharge air temperature − room air temperature = temperature rise Temperature rise = _____	
12	Read total amperage of electric heaters only. Amperage = _____	

Step	Procedure	Check
13	Calculate Btu/hr. Voltage $\times$ Amperage $\times$ 3.414 = Btu/hr Btu/hr = _____	
14	Calculate airflow by the temperature rise method. $$\text{Airflow (CFM)} = \frac{\text{Btu/hr (from step 13)}}{(1.08) \times \text{temperature rise (from step 11)}}$$ CFM = _____	
15	Measure the blower motor amperage. Blower amperage = _____	
16	Rated fan motor amperage from name plate on motor. Rated blower amperage = _____	
17	The measured amperage should be lower than the rated blower amperage.	
18	Turn the thermostat down and observe the heaters come off.	
19	Observe the blower stop 3 min after the heaters come off.	